SACRAMENTO PUBLIC LIBRARY
828 "I" STREET
SACRAMENTO, CA 95814

D1558228

PROPAGATION
TECHNIQUES

PROPAGATION
TECHNIQUES

MITCHELL BEAZLEY

PROPAGATION TECHNIQUES

First published in Great Britain in 2011 by Mitchell Beazley,
an imprint of Octopus Publishing Group Ltd, Endeavour House,
189 Shaftesbury Avenue, London WC2H 8JY
www.octopusbooks.co.uk

An Hachette UK Company
www.hachette.co.uk

Published in association with The Royal Horticultural Society.

This US edition published in 2012.

Distributed in the US by Hachette Book Group USA,
237 Park Avenue, New York, NY 10017, USA

Distributed in Canada by Canadian Manda Group,
165 Dufferin Street, Toronto, Ontario, Canada M6K 3H6
www.octopusbooksusa.com

Design and layout copyright © Octopus Publishing Group Ltd 2011
Text copyright © The Royal Horticultural Society 2011

All rights reserved. No part of this work may be reproduced or utilised in any form or by any means, electronic or mechanical, including photocopying, recording or by any information storage and retrieval system, without the prior written permission of the publishers.

The publishers will be grateful for any information that will assist them in keeping future editions up to date. Although all reasonable care has been taken in the preparation of this book, neither the publishers nor the authors can accept any liability for any consequence arising from the use thereof, or the information contained therein.

ISBN: 978 1 84533 642 4

Set in Gill Sans and Minion
Printed and bound in China

Authors Geoff Hodge, Rosemary Ward
Publisher Lorraine Dickey
Commissioning Editor Helen Griffin
Senior Editor Leanne Bryan
Copy-editor Joanna Chisholm
Proofreader Ruth Baldwin
Indexer Helen Snaith
Art Directors Jonathan Christie, Pene Parker
Senior Art Editor Juliette Norsworthy
Designer Two Associates
Picture Research Manager Giulia Hetherington
Senior Production Controller Lucy Carter
RHS Commissioning Editor Rae Spencer-Jones

CONTENTS

Why Propagate Plants? 6

Rules & Tools 10
  Selecting suitable plant material
  Controlling the environment
  Tools & equipment
  Containers
  Composts
  Watering
  Hygiene, diseases & pests

Cuttings 36
  Introduction to stem cuttings
  Understanding stem cuttings
  Types of stem cuttings
  When to take stem cuttings
  Softwood cuttings
  Semiripe cuttings
  Hardwood cuttings
  Evergreen cuttings
  Heel & mallet cuttings
  Leaf-bud cuttings
  Introduction to leaf cuttings
  Leaf-stalk cuttings
  Midrib cuttings
  Lateral vein cuttings
  Leaf slashing
  Leaf squares
  Monocot leaf cuttings
  Introduction to root cuttings
  Woody shrubs with suckers
  Preparing the parent plant
  Obtaining material & preparing a cutting
  Starting root cuttings
  Tuberous root cuttings
  Introduction to bulbous plants
  Bulbs
  Corms
  Tubers
  Rhizomes

Division 96
  Introduction to division
  Offsets & runners

Layering 102
  Introduction to layering
  Air layering
  Tip layering
  Dropping

Seeds 114
  Introduction to seeds
  Collecting & storing seeds
  Breaking seed dormancy
  Seed sowing in containers
  Caring for seedlings
  Seed sowing troubleshooter

Grafting 134
  Introduction to grafting
  Whip & tongue grafting
  Apical-wedge grafting
  Side-wedge grafting
  Chip-budding

Propagation Directory 146
  Annuals, biennials & bedding plants
  Perennials
  Woody plants
  Indoor plants

Glossary, Index, Picture Credits 188

WHY PROPAGATE PLANTS?

Propagating new plants is fun, interesting and a great way of saving money. There's nothing more satisfying and exciting in gardening than sowing a few seeds and seeing them shoot or taking some cuttings and enabling them to produce roots. Granted, plants started in your home can take longer to produce a display than purchased well-established plants, but home-grown annuals and even some propagated herbaceous perennial plants usually catch up quickly with larger, bought ones.

Seed sowing is an easy way of propagating a wide range of plants – in fact, almost any plant that produces seed. It's even cheaper when you save seed from your own plants. You can also choose from a much wider range of species and cultivars (varieties) than are generally available as bought plants.

Receiving cutting material from a friend's or neighbour's plants or taking cuttings from one of your own plants and getting them to root is basically growing 'plants for free'.

There is no need to worry that you might not have 'green fingers'. These mystical or even mythical digits are simply a result of knowledge – knowing what to do, when and how to do it – all information you will glean in this book.

One term you will need to get accustomed to is 'parent plant'. This refers to any plant that is used to produce offspring. There are two main types of propagation – asexual and sexual.

Asexual or vegetative propagation involves using the plant's ability to produce new organs from adventitious cells; these have the ability to turn into any of the plant's parts – stems, leaves, flowers or roots. Stem cuttings can produce their own roots, root cuttings can develop their own stems and leaf cuttings can create whole new plants. Such vegetative propagation produces young plants that are an exact replica of the parent plant with all the same attributes and characteristics.

Sexual propagation is sowing seed, which is produced as a result of pollination and fertilisation of the female reproductive organs with male pollen. Because sexual reproduction mixes up the genetic material from the two parents, the resulting plants may have slightly different or wildly different characteristics from those of the parents (see box on p119).

Plant up borders more cheaply by propagating all the plants yourself from scratch.

What you can achieve by propagating your own plants depends on how much time, effort and money you are willing to invest. Saving seeds from your own plants and sowing them where you want the new plants to grow is quick, simple and free. Producing enough new plants from cuttings to grow a hedge is fairly straightforward and will take a few months for the plants to root and a few years for them to reach a reasonable height. One quick way of producing a new plant from one that does not readily produce its own roots from cuttings is to graft it by splicing a known cultivar onto a designated rootstock, but this technique does need skill and patience.

It is always best to start with simple subjects, plants and techniques and then move on to more complicated techniques and tricky plants once you have mastered the basics.

One thing to remember is that you will not get a 100 percent success rate. Not every seed will germinate and not every cutting will root, so always do a few more than you actually need. But at the same time do not overdo it. Rest assured, once you have the bug, you will be propagating everything you can lay your hands on. You will also become popular with friends, family and neighbours when you give away excess plants and be someone that the local gardening club looks up to for supplying plants for the club's plant stall or sale.

Spring is a hive of activity for the home propagator, so plan exactly what you want to grow, how many of each you will need and allocate your time accordingly.

Most seeds are sown in spring, but some that need a longer growing season may prefer to be started in late winter. Hardy annuals are sown outside where you want them to grow in late spring, summer and often again in autumn, depending on their hardiness.

It is possible to leave some plants to run to seed and self-sow. However, this can lead to a mass of seedlings, all growing into each other and not developing properly. When this happens the mass of

LEGAL NICETIES

What you propagate is entirely up to you, but there is one area that you will need to be aware of. Plant Breeders' Rights (PBR) are a form of intellectual property legislation designed specifically to protect new cultivars of plants. They allow the breeder to register a plant as their own property if it meets certain relevant internationally agreed criteria. If granted, PBR cover specific territories for defined periods of time. Once PBR have been granted, the owner may license companies to grow their cultivar and collect a royalty on each plant grown. Plants may be protected under UK PBR or under European Plant Variety Rights. Plants with PBR cannot be propagated by any method including seeds, cuttings or micropropagation.

new plants can become 'weeds', getting out of control and swamping more desirable plants. It is far better to collect the seed when it is ripe, sow it yourself and then plant out the young plants at the right spacing. Self-sown seedlings that can become a problem include forget-me-not (*Myosotis*), Shirley and opium poppies (*Papaver*) and even *Helleborus*.

Most cuttings, especially softwood, are taken in spring but this can be done in summer or early autumn. Semi-ripe cuttings are struck in summer and hardwood cuttings in autumn. Root cuttings are taken in winter.

Some plants that can be difficult to propagate vegetatively by traditional methods, or that are badly diseased and traditional propagation would propagate the disease at the same time, are now multiplied by micropropagation or tissue culture. This involves propagating the plant from tiny pieces of plant tissue, often just a few cells, on plates of agar gel. It needs aseptic conditions and is more like chemistry than gardening. In tissue culture, individual or small groups of plant cells are manipulated so they each produce a new plant exactly like the parent. A tiny piece of bud, leaf or stem can produce incredible numbers of new plants in a small space in a short time. Kits are available to try at home, but most gardeners don't have the right conditions, the patience or skill to make this a feasible propagation method.

From bulbs to trees, most plants are easy to propagate at home.

RULES & TOOLS

• SELECTING SUITABLE PLANT MATERIAL •

Knowing how plants grow and develop, which plants are best propagated in what way, understanding what to do, how to do it and when are all important aspects to bear in mind when propagating plants. Using logic and common sense is also essential.

The route to success starts with the selection of suitable plant material and then using an appropriate propagation technique. For instance, some plants are best increased from softwood cuttings, whereas others are best grafted onto a rootstock as they do not root well from cuttings. Some seeds are easy to germinate and if saved from plants in the garden come true to type, looking identical to the plants they were collected from. Others can take months to germinate, often because of an inherent inhibition that may need to be broken first. Seed saved from F1 hybrids or from fruit trees will produce a mixture of varied plants that, as a result of cross-pollination, bear no resemblance to the parent plant they were taken from and so are more than likely to be inferior.

When gathering plant material for propagation, it is essential that you choose carefully. There is no point in selecting diseased plants, especially those affected with viruses, as you will only be passing on the disease. Similarly, do not use weak plants as they will struggle to develop and establish.

When propagating plants with variegated foliage or with unusual characteristics (contorted stems, for instance) always select shoots that exhibit strong variegation or typical characteristics of the plant. However, if you have a plant that has one or two shoots that are different from the rest and that you like the look of, try increasing them. If this change or mutation is stable and continues to show through three or four generations of the plant you have a new cultivar. This is how many new plant cultivars arise.

A plant puts a lot of energy into flower production, whereas you want it to channel that energy into root production, so don't propagate from shoots that are flowering. If necessary, cut back hard one or two of the flowering shoots at the back of the plant, where they will not be missed in the overall display, and then take cuttings from the vigorous regrowth at the appropriate time of year. This vigour will be passed on to the cuttings, ensuring a better chance of success.

When selecting material for cuttings, look for strong, healthy, well-grown stems.

• CONTROLLING THE ENVIRONMENT •

To ensure that propagated material grows well and quickly develops into a strong, healthy plant you may have to control its environment until it becomes established as a new plant.

Two environments are of interest to the plant propagator: the aerial environment around the leaves and stems; and the rooting one in the compost or soil. To ensure good results both environments have to be appropriate for the type of plant material being used.

For successful propagation most plants require an environment that: minimises water loss; provides the correct temperature within the plant's normal range for growth; allows light that is good for photosynthesis but not too strong to cause scorching or excessive drying out; and supports plants in warm, moist compost/soil that provides plenty of aeration and offers the correct amount of nutrients and pH range.

Whereas most hardy plants can tolerate a wide range of environmental conditions, 'softer', half-hardy or non-hardy plants need better-regulated conditions and usually won't survive without some sort of environmental control, especially when propagated and during cooler/colder weather in spring, late autumn and winter.

GLASSHOUSES

Glasshouses provide the ultimate propagating environment, although it can be difficult and expensive to control such a large space. To keep glasshouses frost free over the winter they need some sort of heating, which can be very expensive to maintain at 5–13°C (41–55°F) – temperatures needed for most propagation techniques. It is possible to partition off areas of the

---

SHADING

During strong, sunny conditions the amount of light hitting plants and the temperature under a cover may become critical, causing young plants to wilt or even die. In these circumstances, you'll get better results if you shade the cover to reduce the amount of light transmission.

Shade glass covers by applying a milky shading wash by brush or pressure sprayer. Alternatively, drape shade netting or horticultural fleece over the glass. When using plastic bags (see p17), swap the usual translucent ones for milky ones.

glasshouse with temporary doors or screens of bubble glazing or horticultural fleece or by creating propagating beds on the staging using soil-warming cables or soil-warming blankets. These can be covered with plastic or glass to create heated propagators.

In cold weather, glasshouses can be insulated with bubble wrap to provide 'double glazing', but this does increase humidity and reduce light.

A cold frame can be used for propagation and for hardening off young plants.

COLD FRAMES

A cold frame is a glazed box made either from wood, metal (particularly aluminium) or from brick and glazed with the same materials as a glasshouse. It can be used to propagate a wide range of hardy plants that generally need only the ambient temperature, although small paraffin heaters can be used to keep temperatures higher in spring, autumn and winter.

A cold frame will help to increase the temperature of

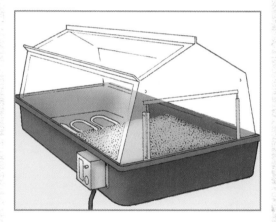

THERMOSTAT-CONTROLLED PROPAGATOR
An electric propagator has a heated base, which provides warmth and better germination and rooting conditions, and a tall, transparent cover.

both air and soil, maintain humidity and ensure good light transmission. One drawback, like many enclosed structures, is that it can overheat during very hot, sunny weather and become very cold when the outside temperature is very low. During hot, sunny conditions, a cold frame can be opened to the elements, which has the disadvantage of reducing humidity, or it can be covered with a shading material, such as horticultural fleece, which also reduces photosynthesis and hence growth.

WINDOWSILL PROPAGATOR
This type of propagator, with individual, miniature seedtrays and covers on a heated base, may be all you need to raise a few tender seedlings and take small numbers of cuttings.

PROPAGATORS

There are two main types of propagator – unheated and heated. The former simply helps keep humidity levels high, whereas a heated one can also produce bottom heat and keep the compost and air temperatures within

HARDENING OFF

A plant moved quickly from a warm, humid environment to a cold or dry one will suffer from shock. To reduce or prevent this shock, acclimatise plants to outdoor conditions over a period of 10–14 days. Such hardening off usually involves moving plants from a heated structure to an unheated, closed one. Then the doors are opened in the day and closed at night; then permanently open but the plants covered with fleece at night; and finally they are exposed to the external conditions with no protection.

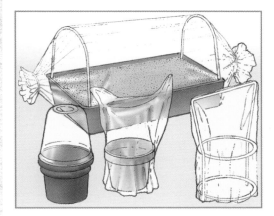

PLASTIC AND POLYTHENE COVERS
An economical way of making propagators is to use plastic or polythene covers. Keep the cover away from the foliage of the plants.

MIST PROPAGATION

For those gardeners serious about increasing their stock of plants or for those regularly propagating from plants that are difficult to root from cuttings, a mist propagation unit is a good investment. This has an 'electronic leaf' that senses drops in humidity and rectifies this. The warm, damp atmosphere created by mist propagation units prevents cuttings from wilting and scorching and encourages rapid root growth, even with normally difficult subjects.

PLASTIC COVERS

When propagating on a budget, a variety of plastic covers can be used to cover propagation frames or individual pots.

Polythene bags can be placed over trays or pots or supported on three or four split bamboo canes or wire hoops inserted into the compost and kept in place with string or elastic bands.

Alternatively, cut the bottom from a 1- or 2-litre plastic bottle, which fits perfectly over a 9cm (3½in) pot.

higher and more suitable for good germination and rooting.

Some heated propagators are thermostatically controlled, which means the temperature can be regulated to the optimum for germination and rooting. Those without a thermostat produce a standard heat output and often promote critically high temperatures in warm, sunny conditions. Thermostat-controlled propagators offer the most versatility and are good choices for those taking their propagating seriously.

• TOOLS & EQUIPMENT •

For successful propagation you will need a selection of tools and equipment to help you get the best results. The following will assist your choice.

● The most important piece of equipment you'll need for taking cuttings is a grafting, pruning or budding knife. Aim to keep it for propagation only. For

BASIC TOOLKIT

Potting/propagation bench, propagation area
Propagation knife, modelling knife or razor blades
Secateurs
Sharpening/honing or carborundum stone and oil or diamond sharpening tool & cleaning cloths and liquids
Sieve (3mm/$\frac{1}{8}$in & 6mm/$\frac{1}{4}$in mesh)
Compost tampers, dibber & widger
Pot labels & pencil/marker pen
Small polythene bags & elastic bands

Split bamboo canes
Hand sprayer/mister
Small watering can
Hormone rooting powder or gel
Horticultural fleece
Glass panes &/or clear plastic sheets
Cuttings board or pane of glass
Small pots, seedtrays & other containers (see p22)
Propagators (see p16)
Composts & compost scoop (see p26)
Fungicides & insecticides (see p32)

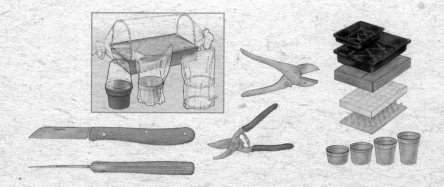

RULES & TOOLS

most propagation methods you need a medium-weight knife with a carbon steel blade, but for grafting a heavier knife is more suitable. A budding knife has a spatula end for prising open flaps of bark. Even though all blades should be handled with care, a few plasters are always worth adding to the toolkit.

- A good pair of secateurs is important for removing cuttings material from the parent plant, preparing hardwood cuttings and when grafting. When choosing a pair of secateurs pick them up and see how they feel in the hand. Make sure they're not too heavy and that the spread of the handles when open suits the size of your hand. A spring-loaded return action will make using them much quicker. Some models have a sap groove, which helps reduce the build-up of sticky sap and so prevents the blades sticking. As with knives, spend as much as you can afford and always look for a good-quality steel blade that can be sharpened easily.

- Regularly use the sharpening/honing or carborundum stone lubricated with general oil to sharpen/hone blades, or do this job with a diamond sharpening tool. A blunt blade will make ragged cuts, which can be a source of disease and when used on cuttings can prevent good rooting. All cutting tools must also be clean. A dirty blade may hold diseases that can be spread from plant to plant, so sterilise the blade regularly with household bleach or Jeyes Fluid. Wipe after daily use with oil.

- A small-mesh sieve is useful for sieving large lumps of compost when a fine compost is needed – for small seeds, for example.

- Use compost tampers to compress compost lightly and to help create a flat, level sowing surface. A pot or tray of the same size can be used instead. A dibber is good for making holes in the compost for the cuttings and a widger for lifting seedlings when pricking out and potting on.

- Pot labels and pencils or marker pens are essential so that you know what is what in each pot or tray.

- Put cuttings in small polythene bags after collecting them to help prevent them drying out. The bags can also be used to cover pots and trays as cheap propagators. Elastic bands help keep

CHOOSING A KNIFE

In most circumstances a knife with a straight cutting blade is the best choice, especially as it is easy to sharpen. Another point to look for is a full-length tang – that is, the steel extends the full length of the handle to give the knife better strength and stability during use. The back end of the blade, or shoulder, should also be set well back into the handle when opened. A general-purpose gardening knife has a rounded tip (see above), whereas the tip of a budding knife is pointed.

polythene bags over pots and seedtrays in place and seal them.

● Split bamboo canes support polythene bags and keep them away from cuttings and developing seedlings.

● Use the hand sprayer/mister to keep cuttings and compost moist with water or a fungicide.

● The small watering can is for watering the compost; make sure it has a fine rose.

● Hormone rooting powder or gel is useful to help encourage rooting.

CUTTING HARDWOOD

Right-handers should hold the plant in the left hand. With the blade positioned below the stem and right thumb above, make a shallow-angled cut from beneath, drawing your right forearm backwards and maintaining the gap between right thumb and blade. Done in this way, protective gloves are not needed. In other circumstances protect your hands when cutting hardwood.

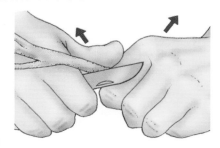

SHARPENING A FLAT-GROUND BLADE

Draw along the sharpening stone to the end with slight pressure on the sharpened edge. Repeat as needed.

SHARPENING A HOLLOW-GROUND BLADE

Hold the blade at one end of the sharpening stone, facing forward at an acute angle to the stone surface.

Push gently along the length of the stone, maintaining the angle between blade and stone.

- Horticultural fleece will insulate against cold conditions and provide shade when it is very sunny.

- Use small panes of glass &/or clear plastic sheets to cover trays or pots in order to increase humidity and maintain warmth around plant material.

- Cut softwood plant material on the cuttings board or pane of glass.

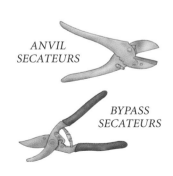

ANVIL SECATEURS

BYPASS SECATEURS

SECATEUR TYPES

There are two main types of secateurs – bypass and anvil. Anvil secateurs have a blade that cuts against a flat anvil plate and generally crush the stem while they are cutting, so are not suitable in propagation. Bypass secateurs cut with a scissor action, producing much cleaner cuts, and so are much better for taking and preparing cuttings material.

• CONTAINERS •

For the vast majority of propagation techniques you will need a selection of pots, trays and other containers in which rooting or germination can take place.

POTS

Pots are probably the most versatile containers for seed sowing, potting up and taking cuttings.

Plastic pots are a better choice than clay ones as they do not dry out as quickly and so the compost remains at the right moisture level for longer. Clay pots are often better for plants that don't like to be kept too wet, such as cacti and succulents. Plastic pots are also cheaper, lighter and more durable than their clay counterparts and are easy to clean.

For most propagation needs you will require only three or four different sizes of pots. The most common size is 9–10cm (3½–4in) diameter. Make sure there are plenty of drainage holes in each pot – especially important if the compost is to be kept moist by standing the pots on dampened capillary matting.

Standard pots are as deep as they are wide and the most commonly used, while three-quarter or half pots (referred to as pans) are good for seed sowing. Long pots, referred to as long toms, are suitable for long tree or shrub cuttings, especially hardwood ones.

Although most pots are round, square ones can be packed closer together. They also have a larger capacity: a 7cm (2½in) square pot contains the same amount of compost as a 9cm (3½in) round pot. It is more difficult to compress compost lightly in square pots, so a square pot tamper is a useful tool.

POTS AND CONTAINERS
You will need a range of pots, trays, celltrays and other containers for different plants and propagation methods.

Disposable pots

It is also possible to buy various disposable pots in a number of sizes that are made from organic materials, such as compressed peat or coir. These have the advantage that the plant roots will grow through the sides and the whole thing can be transplanted or planted in its entirety without disturbing the roots. The watering of these is usually trickier than plastic pots, since if they dry out, the roots may be damaged or killed; if kept too wet, disposable pots tend to disintegrate early.

Pellets & blocks

Like disposable pots, pellets and blocks have the advantage of eliminating root disturbance when transplanting or planting out, as the roots grow right through to the sides – a visible indication that the plant is large enough to move on. These compost pellets need careful watering for good results.

The compressed peat or coir is held in a net and comes as a flat disc when dry, expanding into a pellet when thoroughly moistened. A cutting or seeds can then be placed in the depression at the top of the resulting block.

SEEDTRAYS & CELLTRAYS

In some situations, especially when a large number of seedlings are needed, a seedtray is better than a pot. Being shallow, a tray is suitable for only the smallest of cuttings. Where fewer numbers are needed, half trays or even quarter trays are available.

Nearly all seedtrays are now made from plastic – wooden trays being rare. As with pots, plastic trays are lighter, cheaper and easier to clean and reuse

DISPOSABLE POTS
Plant roots grow through the sides of peat or coir pots, provided the pots and potting compost are kept moist.

MAKE YOUR OWN SOIL BLOCKS & POTS

Soil blocks
You'll need to buy or even make a steel blocking tool, which is needed to create the square compost blocks. Bought versions have a plunger that pushes out the compost block and creates a small hollow in the top for the seed or cutting. The blocks are prone to drying out quickly and require careful watering if they are to work properly.

For soil blocks the compost needs to be quite stiff, so that it holds together. There are numerous compost recipes available to make your own blocking compost – just search the internet for options. One recipe uses four parts moss peat sifted through a 6mm (¼in) sieve, mixed with one part home-made compost (or worm castings), lime and rock dust. Mix and thoroughly wet the ingredients until they are stiff and moist, but not soupy or too dry.

Paper pots
You can also make your own pots from strips of old newspapers or magazines. Several mail order companies supply a simple wooden tool/former that will produce a pot in minutes. Such pots are sufficiently strong to hold together while the seedlings are establishing, but quickly break down when planted in the ground. Another considerbale benefit is that you can make as many pots as you require, when you need them, and there is no root disturbance when transplanting.

than wood, although flimsy plastic may not last for more than one season.

Celltrays are becoming ever more popular for seed sowing and rooting small cuttings. These are made from plastic or polystyrene (which provides extra insulation and heat retention – important when using a heated propagator in early spring or autumn).

The advantage of using celltrays over standard seedtrays is that each plant in each cell develops its own rootball, so there is no root disturbance when transplanting or planting out.

Keep down costs by reusing thoroughly cleaned plastic pots and polythene bags.

REUSE & RECYCLE

Before reusing containers carefully clean out any old compost and other deposits with a soft brush or sponge. Then, wearing rubber gloves, sterilise the containers in a bucket containing a mild solution of household bleach. Clay pots will need to be soaked longer than plastic ones. Rinse well (especially clay pots) and dry before storing. Many gardeners recycle old yogurt pots, plastic vending-machine cups or fruit and meat punnets. These should also be thoroughly cleaned out and sterilised before use and drainage holes made in the bottom. The easiest way to do this is to burn the holes carefully with a hot pin or nail. Toilet-roll inner tubes are the perfect choice when sowing seeds that need to develop a long root, such as sweet peas, peas and beans.

• COMPOSTS •

The term compost can cause some confusion, as it refers to two quite different things: potting media used in containers; and garden compost. The former (called just 'compost' hereafter) comprises potting, seed sowing, multipurpose and cuttings composts, all of which have specific mixes of ingredients, often to set and prescribed recipes, and are used for specific plant growing purposes. Home-made garden compost refers to material produced by rotting down garden and kitchen waste, usually in a compost heap or bin.

Composts need to keep moisture close to the seed or base of the cutting; they have a fine but open, well-drained texture that holds air; and they retain warmth. Their main constituent is an organic medium, such as peat, coir or composted bark, which holds the moisture. Sharp or gritty sand or grit, perlite or vermiculite is used to keep the compost open and so it has tiny air spaces – essential for good root growth as roots need to breathe.

Composts may or may not contain nutrients – and how much depends on the compost type, which relates to the propagating material for which it is intended. Seeds need a very small amount of nutrients, while cuttings need none. Young plants require some nutrients, but not as much as established plants, so their compost contains fewer nutrients than potting or multipurpose compost.

Seed and potting composts also contain lime, but these should not be used for plants that prefer to grow in an acid medium (pH 6.5 or lower). Such acid-loving or lime-hating plants require ericaceous compost.

Although many gardeners buy multipurpose compost for all their compost needs, the avid propagator will have a selection of different ones for varied uses. At the minimum you will need a seed sowing compost, a young plant compost and a cuttings compost.

It is worth sticking to a compost mix/recipe once you are used to it. Being familiar with a particular mix will ensure you get the best results, because you know how much water to add, how quickly it dries out and so on.

MAKING YOUR OWN COMPOSTS

If proprietary composts do not provide the exact growing conditions you need, you may decide to make your own. However, before doing this, experiment with your familiar mixes first. All composts 'work' differently and you might find some minor adjustment is all that is required. If this does not provide success, a quick trawl of the internet will reveal numerous recipes. For John Innes No 2 use twice the amount of John Innes Fertiliser Base compared with John Innes No 1 (see box below); for John Innes No 3 use three times.

HOME-MADE COMPOSTS

Seed sowing compost
(parts by volume for 36 litres compost)
2 parts sieved peat, coir or composted bark
2 parts sharp sand, vermiculite or perlite
1 part sterilised loam
40g (1½oz) ground limestone and 20g (¾oz) superphosphate

Cuttings compost
Equal* parts sieved organic medium – peat, coir or composted bark – and sharp, gritty sand, vermiculite or perlite.
* For plants that prefer a drier medium, such as pelargoniums and alpines, use up to one-quarter organic medium and three-quarters sand, vermiculite or perlite.

Young plant potting composts (John Innes No 1)
(parts by volume for 36 litres compost)
7 parts sterilised loam
3 parts sieved peat, coir or composted bark
2 parts sharp sand, vermiculite or perlite
20g (¾oz) ground limestone and 120g (4oz) John Innes Fertiliser Base

Loam-free potting compost
(parts by volume for 36 litres compost)
3 parts sieved peat, coir or composted bark
1 part sharp sand, vermiculite or perlite
120g (4oz) each ground limestone and John Innes Fertiliser Base

ENCOURAGING THE DEVELOPMENT OF ROOTS

Dip the base of a stem cutting into water so hormone rooting powder will readily adhere to the cut surface at the base of the cutting.

Push just the base of the cutting onto the hormone powder. Ensure no hormone powder adheres to the outside of the stem cutting.

ROOTING CUTTINGS IN WATER OR GEL

It is also possible to root cuttings in jars of water, which is a cheap and convenient way of producing new plants, or you can do this in rooting gels that have added hormones, nutrients and minerals to aid fast rooting of cuttings. In both methods you can see the roots form and so know when it is time to pot up the plants.

HORMONE ROOTING PRODUCTS

Some cuttings are likely to take a long time to root or otherwise be difficult, but using a hormone rooting powder or gel may improve rooting success. However, they are not miracle cures and are no substitute for correct practices.

The hormone is very volatile and soon evaporates, so always ensure you have fresh stocks available – preferably buying it annually. Keep it cool and dark in the fridge, to ensure the hormone stays as fresh as possible.

Only small quantities of the hormone are needed, so dip only the end of the cutting in the powder and immediately tap off any excess. If you use too much, the hormone will actually inhibit rather than encourage rooting.

To avoid possibly contaminating the whole lot, tip a little hormone out into the lid or onto a saucer; dispose of this after use rather than returning it to the original container.

A glasshouse is a propagator's best friend and enables a wide range of plants to be increased.

· WATERING ·

To ensure good seed germination and fast rooting of cuttings, there should be adequate amounts of moisture in the compost – but not too much to cause waterlogging and subsequent rotting.

When watering, always use tap water rather than water from a water butt as the latter may contain pathogens that can cause problems with seeds and young plants. Always use tepid water, as cold water can shock young plants.

Sow seeds in a slightly dry compost and carefully water them afterwards. When watering from above always apply enough water so that it just begins to trickle from the drainage holes at the bottom of the pot. For fine seeds and delicate seedlings it's usually better to water from below. After sowing the seeds, stand the pots or trays in a bowl or sink filled with 5–7.5cm (2–3in) of tepid water. Remove them when moisture just begins to break the surface of the compost. Then drain before moving to their germination area.

Where high moisture levels need to be maintained stand the containers on capillary matting with one end placed in a container of water.

When propagating from cuttings you can water the compost before placing it in the propagation container. To establish the right moisture level, gently squeeze a handful of compost. If it falls apart, it is too dry. If water trickles out of the compost, it is too wet. The

RECOMMENDED WATERING TECHNIQUES

Use a watering can with a fine rose. Start by pouring the water away from the container. Once there is an even flow, direct the water over the compost. Move the can away before stopping the flow of water.

To water by capillary action, stand the containers in a shallow bath of water. When the compost is thoroughly wet, remove the containers and leave to drain.

correct moisture content is when, after squeezing, the compost holds together in one piece and little beads of water form on the outside.

NATURAL FOOD RESERVES

Seeds store within themselves all the food and nutrients they need during germination and the early phases of growth – the bigger the seed, the greater the food store. Therefore tiny seeds may need transplanting quickly from their seed compost to a young plant one (see p27) or else may benefit from careful feeding soon after germination. Be very sparing in this as too much fertiliser can cause any remaining seeds to rot and produce stunted seedlings.

Cuttings should already contain enough food and nutrients for rooting and early growth because plants store food reserves and nutrients within their stems, leaves and roots. As with seeds, the bigger the cutting, the more food is stored and so the longer it can survive before rooting and growing on. Therefore providing you use the appropriate compost you should not need to add extra fertiliser until the plants have been potted on or planted out. In fact additional fertiliser at an early stage can actually inhibit rooting of cuttings.

Always prick out, transplant or plant out into a richer compost once the plants are large enough to handle

Seeds store enough food reserves for the early stages of growth.

and certainly before they start to show signs of nutrient deficiency. Once established garden plants will need regular feeding to provide sufficient nutrients.

• HYGIENE, DISEASES & PESTS •

Seeds, seedlings and cuttings, especially those that have not yet rooted, are very susceptible to problems from pests, and even more so from diseases. It is not until they become established plants that they really build up any resistance to these hazards. This means that good hygiene is the key to success.

Always ensure that your propagation area or potting bench is clean and tidy and all tools, propagation materials and composts are clean and preferably sterile – especially cutting tools.

DISEASES

Successful propagation starts with selecting healthy plant material. Never use diseased plants because the disease will be passed on to the new ones. Viruses and bacterial diseases readily spread from plant to plant and are often the most difficult to spot – so be vigilant.

To ensure they are disease free treat most leafy cuttings with a systemic or copper-based fungicide and spray again once they have been inserted into the cuttings compost and before they are

Clean conditions are vital for healthy plants, so be scrupulous about hygiene.

Damping-off disease causes young seedlings to collapse and fall over.

placed in the propagator. Very high humidity in the propagator can quickly lead to problems with grey mould (botrytis), which rapidly spreads and kills or severely weakens plants. Therefore improve ventilation to regulate humidity and use the fungicide again if conditions are too damp and you suspect there may be a problem.

Cuttings are prone to basal rot and young seedlings to damping-off disease, especially if sown too densely or in unsterile or spent compost or if watered with water stored in a water butt. Always use clean containers and fresh compost for seed sowing and sow thinly over the surface of the compost. If you are concerned about damping off, protect the plants with a copper-based fungicide.

Prick out young seedlings as soon as they are well developed and large enough to handle and move. This way you can reduce any possible problems with damping off and other post-germination problems.

TIDEMARKS ON CONTAINERS
All possible sources of infection must be eradicated, so remove any crusty layers of soil or chemicals on the sides of pots and seedtrays. Clay pots should also be soaked to ensure they are clean.

Always check for pest and disease problems when handling plants.

Check your young plants on a regular basis and ensure that you deal with any problems that you discover quickly before they get out of hand.

PESTS

Uncovered seedlings and young plants can be attacked by the whole range of plant pests, so check regularly and deal with any outbreaks as soon as they occur. Pay particular attention to stems and the underside of leaves, which is where most pests congregate. If you suspect pests you can spray with an insecticide.

Adult sciarid or compost flies are often a nuisance flying around pots of compost. However, it is the larvae that cause more concern as they eat plant roots. Sciarid damage is always worse in peat- or coir-based composts and especially in composts that are kept too damp. Unless there is a large population of larvae in the soil the actual damage is usually minimal, but it may still be worth treating the compost with systemic soil-drench insecticides or even a biological control based on nematodes (*Steinernema feltiae*).

Aphids, whitefly and red spider mite will often congregate on young plants, sucking sap and reducing vigour. Red spider mite is always worse in hot, dry conditions; high humidity will stop them getting out of control.

PESTS & DISEASES

*Sciarid fly & grub
(fly about 6 times
actual size; grub
about 4 times
actual size)*

*Spider
mite
web*

*Adult glasshouse whitefly & pupa
(fly and pupa about 12 times actual size)*

*Slug &
damaged
leaf*

Basal rot in cutting

If you learn to identify the most common pests and diseases that affect newly propagated material and inspect your plants regularly, you can nip any attack in the bud.

CUTTINGS

• INTRODUCTION TO STEM CUTTINGS •

Growing new plants from short sections of cut stems is the most useful and most versatile method of vegetative propagation. It is suitable for increasing your stock of trees, shrubs, climbers, roses, conifers, herbaceous perennials, fruit, herbs, indoor plants and half-hardy perennial bedding plants. It is also a quick process, as some plants root very easily, often in a matter of weeks.

The aim when propagating from stem cuttings is to initiate and develop a good root system on the stem from their adventitious cells. The processes controlling root initiation are essentially chemical: the higher the temperature, the faster the chemical reaction and thus root production.

Always remember that, no matter how good you are, not every cutting will root successfully, so always take a few more than you actually need.

IMPROVING ROOTING SUCCESS

Some plants are more difficult to root than others and you may need to take different types of cuttings at various times of years to see which one is the most successful. It is also possible to speed up and assist rooting by trying a number of other approaches such as applying a hormone rooting powder or gel (see p29).

Some hard-to-root plants respond well to wounding the stem. In this a thin, vertical sliver of bark, up to 2.5cm (1in) long, is removed at the base of the cutting and the wound is then dipped in hormone rooting compound.

Cuttings of large-leaved plants – hydrangea, for example – may be more successful if their leaves are cut in half horizontally, which reduces water loss.

Other plants root better when removed from the parent plant with a heel (see p56) or when taken as a mallet cutting (see p57).

A lot of gardeners overfirm the compost when filling their containers, and this drives out air, which is essential for good rooting. For best results, loosely fill the container with compost, lightly firm with two fingers, add more compost to the top of the pot, tap the pot to settle it and then lightly firm the compost to within 1cm (½in) of the rim.

Although dibbers are often used to make holes in the compost for cuttings, these can be too large at the base of the cutting, and this prevents rooting. If this is a problem, gently push in the cutting until its base is surrounded by compost.

PARTS OF A STEM

As different cuttings methods use different sections of stem, it is worth knowing the names of stem parts. At the tip of the shoot is the apical bud from where the stem grows. Below it are the leaves, which are joined to the stem at an axil – the angle between the leaf and the stem. In each axil is an axillary bud, which produces branching shoots or sideshoots. Where the leaves and axillary buds attach to the stem is a node. The section of stem between two nodes is an internode.

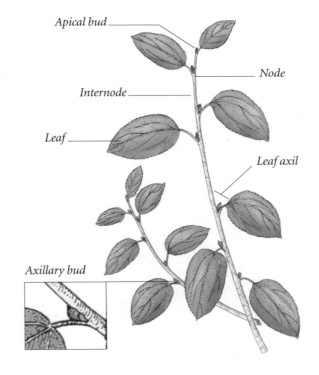

• UNDERSTANDING STEM CUTTINGS •

Taking stem cuttings is easy – providing you have a good sharp knife and/or secateurs. The challenge is keeping them alive and healthy during the rooting process. Depending on the type of cutting used, drying out and fungal rots are the main problems to overcome.

Until the cutting has rooted, it relies on the food reserves and nutrients stored within its stem and leaves, plus food produced by the leaves (if present) through photosynthesis. This has an important bearing on the age and size of the stem used as a cutting, since the bigger the cutting (both length and width), the more food is stored and so the longer it can survive before rooting.

Similarly, the more leaves it has, the more food it can generate. However, as plants lose water through pores on the underside of their leaves, the greater the leaf area, the more water the cutting will lose and so the greater the risk of drying out before rooting. This means you have to strike a balance between the size of the cutting used and its total leaf area left behind after you have prepared it.

Immature (softwood) cuttings contain the lowest food reserves and are at the greatest risk of both drying out and rotting before rooting. This means you need to get them to root as quickly as possible by controlling the environment with heated propagators and close-fitting covers or similar protection. Hardwood cuttings, on the other hand, are more mature, have greater food reserves and are less prone to drying out and rotting, so can be left to root more slowly with less environmental protection.

To root well cuttings need a delicate balance of heat/warmth (particularly in the rooting medium), bright but indirect light, high humidity and adequate ventilation. If the environment is too warm and/or too dark, it will force the cuttings to grow tall and spindly.

OLD STEMS

Young stems have more potential to root than older ones and where a plant has only old wood it should be cut back hard to encourage new vigorous shoots. However, if the plant is very old, even these new shoots will root less readily than stems from a young plant.

Pelargoniums are quick and easy to propagate from stem cuttings.

• TYPES OF STEM CUTTINGS •

There are four main types of stem cuttings – softwood, greenwood, semiripe and hardwood. They are taken at different times of the year, depending on when the necessary plant material becomes available.

When propagating plants that are difficult to root, stem cuttings can sometimes be further modified by taking basal cuttings (see p44), heel cuttings (see p56), mallet cuttings (see p57) or leaf-bud cuttings (see p58).

Evergreens can be treated by any of the methods mentioned.

● Softwood cuttings are taken from the young flush of new growth – usually in spring. They tend to root very quickly, but need a carefully regulated environment to minimise water loss, drying out and rotting (see p44).

● Greenwood cuttings are similar to softwood cuttings, but are usually taken from late spring to midsummer. Their stems are soft and so need a protected environment (see p47).

● Semiripe cuttings are taken from midsummer to midautumn from stems that have started to harden. They are more resilient than softwood cuttings (see p48).

● Hardwood cuttings are taken from the leafless, dormant stems of deciduous plants from midautumn to late winter. They are the hardiest type (see p50).

SOFTWOOD CUTTINGS   GREENWOOD CUTTING   SEMIRIPE CUTTING   HARDWOOD CUTTING

• WHEN TO TAKE STEM CUTTINGS •

Although some plants should be propagated at a specific time of year using a specific method, there are lots of plants that can be rooted from cuttings at just about any time of year. If the plant has produced sufficient suitable growth, it makes sense to take the cutting and try to root it.

WINTER TIMING

The critical phase for cuttings taken in late summer or early autumn, especially softwood cuttings of half-hardy perennials, is during winter. To get them through the cold weather successfully, they need to have developed a good root system. To help cuttings survive if they haven't produced a large enough rootball by autumn you should leave them in their rooting container and pot them up only in the following spring.

WHERE TO CUT A STEM

The vast majority of stem cuttings have their basal cut made immediately below the swelling just below an axillary bud and leaf – at the node. Such nodal cuttings contain a preponderance of cells that have the ability to form roots and a concentration of hormones to stimulate root production.

Some plants also root well when the basal cut on their more mature stems is made some way below a node (at an internode). Such internodal cuttings are stronger and less likely to rot.

WHERE TO CUT A STEM – NODAL AND INTERNODAL CUTTINGS

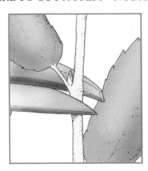

A nodal cutting has its basal cut just 3mm (⅛in) below a leaf joint, or node. It is the traditional place to cut soft, immature stems just below a node these are harder and more resistant to fungal rots than if cut further away.

An internodal cutting has its stem cut further away from the leaf joint, or node, so it is between nodes. This type of cutting is used mainly for more mature, woodier stems, and you can obtain more cuttings from each stem.

• SOFTWOOD CUTTINGS •

Softwood cuttings are a popular way of propagating numerous trees, shrubs and perennials, half-hardy bedding plants and even houseplants.

Soft wood, or growth, is the most immature part of the stem, produced continuously during the growing season at the tips of the stem. Although softwood cuttings can be taken at any time during the growing season, they are more commonly struck in spring. They can also be removed in summer and even early autumn from stems that have been cut back hard to encourage strong new growth.

Because of their softness, these cuttings are often the most difficult to keep alive and to prevent from wilting or rotting. Fortunately, they have the greatest potential of all stem types to produce roots, because growth is fastest when buds first break in spring. Pot-grown plants that have not produced enough growth to produce sufficient cutting material can even be forced into growth by moving them to a glasshouse or other warm place at a temperature of 25–30°C (77–86°F).

The rapid growth rate of softwood cuttings does have one drawback – they lose a lot of water. Cuttings should

A wide range of perennials – both hardy and tender – are easy to propagate from softwood cuttings.

therefore be taken in the morning when they are full of water (turgid) to avoid excessive wilting. Once they have started to dry out and wilt cuttings will never root, so it is vital that they are protected against extensive water loss. Collect only enough material that you can deal with immediately and place it at once in a bucket of water that is sealed to maintain humidity. A piece of damp

BASAL CUTTINGS

Select clusters of young, sturdy shoots, 8–10cm (3–4in) long, with the leaves just unfolding. Using a sharp knife remove them as close to the base as possible, including part of the woody basal tissue. Then treat as softwood cuttings (see p46). This technique is suitable for many herbaceous plants in spring, such as asters, chrysanthemums, delphiniums, lupins, phlox and salvia.

cotton wool placed in the bag will help reduce water loss. If the cutting material cannot be prepared immediately, place the bag in the salad drawer of a fridge to slow down metabolism; cuttings can be kept here for up to 24 hours, but it is far better to use them well before then.

Insert each prepared cutting so that its bottom leaves are in contact with the compost and space them so their leaves

PROPAGATING FROM SOFTWOOD CUTTINGS

Fill a container with moist cuttings compost. Lightly firm to within 1cm (½in) of the rim.

Preferably in the early morning remove the top of a healthy, non-flowering shoot.

Put the cutting at once in a polythene bag or a bucket of water and place in the shade.

Shorten each cutting to 5–10cm (2–4in) long, cutting just below a leaf joint (node) with a sharp knife.

Remove the leaves from the lower third of the cutting. Dip the basal cut in a powder fungicide.

Make a hole with a dibber in the compost. Insert each cutting up to its lower leaves.

do not touch. Mist the cuttings with a fungicide. Place the pot in a propagator with bottom heat of 18–24°C (65–75°F); alternatively cover with a plastic bag, supported on a wire hoop, seal around the pot with an elastic band and place it somewhere warm.

Ensure the compost is kept moist until the cuttings are well rooted. At least twice a week ventilate the cutting for 10 minutes. Remove any dead, rotting, dying or diseased material at once.

Once they have rooted, harden off the cuttings for about two weeks (see p16) before potting them on individually.

GREENWOOD CUTTINGS

These are taken later than softwood cuttings, usually from late spring to midsummer. They are similar to softwood cuttings, except that the base of the stem is firmer as it has had longer to mature.

Prepare greenwood cuttings as you would softwood cuttings, but make them longer, generally 8–13cm (3–5in) long. This technique is suitable for ceanothus, forsythia and philadelphus.

Settle in the cuttings by watering. Label, then place in a propagator positioned out of full sun.

Once they have fully rooted, harden off the cuttings gradually, keeping the compost moist.

Pot up the cuttings individually in small pots of potting compost, and water in to settle the compost.

• SEMIRIPE CUTTINGS •

Semiripe cuttings are taken from midsummer until midautumn, when growth slows down and stems start to harden up. The base of the cutting should be hard, while the tip is still soft and the stems should preferably have short internodes (gaps between the buds/leaves).

As they are thicker and harder than softwood cuttings and have a greater store of food reserves, semiripe cuttings are much easier to keep alive. They will usually root in even low light levels. However, they have a similar vulnerability to water loss and wilting as softwood cuttings because they both carry leaves.

Semiripe cuttings taken in summer do not need bottom heat, but they do require good humidity. Therefore they are rooted in soil in a covered cold frame, low polythene tunnel or similar structure. Dig over the soil and add organic matter and grit to improve drainage. Rake level and lightly firm the soil, then cover with 2.5–4cm (1–1½in) of fine sand.

When only a few plants are needed, semiripe cuttings can be rooted in a pot of cuttings compost. Place the pot in an unheated propagator or cover it with a milky polythene bag. To keep the bag away from the leaves, hold it up on a wire hoop or by a support made from split bamboo canes, then seal around the pot with an elastic band. Place the pot in a warm, light position, but out of direct sunlight, in an unheated glasshouse or a cold frame or on a windowsill. If this isn't possible, then shade the plants from direct sunlight by draping a layer of horticultural fleece over the pot or propagator.

Semiripe cuttings mainly root in late winter and spring, but some may develop roots in a mild autumn. Those taken in autumn usually root more quickly in a heated propagator. Cuttings in a cold frame may not properly root until late spring. In which case, it is important to insulate the cuttings and frame against severe weather and frost. Horticultural fleece draped directly over the cuttings will give about 5°C (7°F) of protection; covering the frame with carpet, bubble glazing or similar protection will help protect against lower temperatures. Always remove the insulation when the temperatures rise, especially during the day, so plants have access to daylight. Otherwise cuttings can become drawn and leggy.

BASAL CUTTINGS

Basal cuttings are where the shoot is severed at its base, including part of the woody basal tissue and the slight swelling here. This method of propagation is frequently used for brooms – *Cytisus* and *Genista*.

PROPAGATING FROM SEMIRIPE CUTTINGS

Prune the parent plant at the start of the dormant season to encourage strong stems to grow.

Select cuttings 10–15cm (4–6in) long. Remove any soft tips and all leaves from the bottom third.

Insert each cutting into the soil in the frame so that the bottom leaves are in contact with the sand covering.

Space the cuttings 7.5–10cm (3–4in) apart. Label and water. Then close and shade the frame.

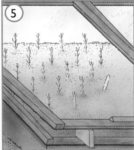

Once the danger of frost is over remove any insulation and air the frame. Apply a liquid feed.

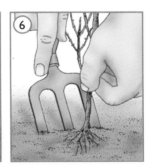

After the leaves have dropped in autumn lift the new plants and transplant them. Label them clearly.

• HARDWOOD CUTTINGS •

Taking hardwood cuttings is the easiest stem cuttings method as they have no leaves to rot and plenty of stored food reserves. It is perfect for a wide range of deciduous trees, shrubs, roses and soft fruit. Although this type of cutting may be slow to develop roots and shoots, it is usually successful. To increase your chances of success, cut back the parent plant in spring and use the more vigorous regrowth as the cutting material.

Hardwood cuttings are taken when the plant is fully dormant after leaf fall in autumn and winter, avoiding periods of severe frost; the optimum time is usually midautumn. Since the cuttings have no leaves at this time, there is little needed in the way of environmental control, so they are usually grown outdoors in the ground in a prepared 'slit' trench.

Ensure the soil is well dug over and additional organic matter added if the soil is light sand, heavy clay or otherwise in poor condition. To make the trench, insert the blade of a spade vertically in the soil and push it forward to produce a straight-sided, V-shaped trench. Add 2.5–4cm (1–1½in) of sharp or gritty sand in the bottom.

You can also grow a few hardwood cuttings in a container.

HARDWOOD CUTTINGS IN CONTAINERS

If you are taking only a few cuttings, there is no need to dig a trench; simply insert the cuttings into deep containers of gritty potting compost, such as a 50:50 mix of coarse grit and multipurpose compost. Keep the pots in a sheltered cold frame or unheated glasshouse until the following autumn, ensuring that they do not dry out.

SLOW-ROOTING PLANTS

If outdoor conditions are unsuitable, or if plants are either slow rooting – dogwoods and laburnum, for example – or difficult to root, plant bundles of 10–12 hardwood cuttings in well-prepared soil in a cold frame or pot filled with moist sand until spring. In early spring, before the leaf buds break, make a trench and set out the cuttings as described opposite.

Select and prepare vigorous, healthy shoots that have grown in the current year. The sloping cut above a bud at the top of each hardwood cutting may help to shed water and acts as a reminder of which end is the top. Insert the cuttings into the slit trench with two-thirds of the cutting below ground level. The few buds remaining above the ground will form the new branch structure of

PROPAGATING FROM HARDWOOD CUTTINGS

In late autumn remove a suitable hardwood cutting with all its current year's growth.

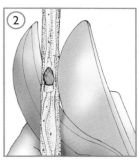

Make sure the proposed tip of the cutting is hard wood, then make a sloping cut just above the top bud.

Make a horizontal cut below a bud to produce a cutting 15–30cm (6–12in) long.

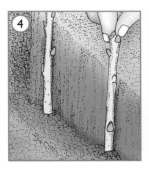

Treat with rooting hormone, then insert in a slit trench with a layer of sharp or gritty sand in the bottom.

Backfill the soil and firm, leaving 2.5cm (1in) of the cuttings exposed. Label and water in.

During the following autumn lift each rooted cutting and transplant to its final position.

Dogwoods, grown for their colourful stems, are best propagated from hardwood cuttings.

the developing plant. Where a single-stemmed plant is needed, such as with gooseberries or poplars (*Populus*), leave only one bud at ground level (see box below) or rub off surplus buds.

Allow 10–15cm (4–6in) between cuttings and, if a second row is needed, 45cm (18in) between trenches.

Although most plants need no protection, some, such as dogwood (*Cornus*) benefit from protection with a cloche or cold frame.

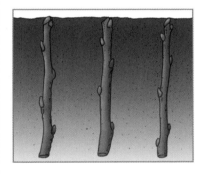

SINGLE STEMS
To grow single-stemmed plants (for example, standard trees) rather than shrubby ones with several stems, you need to encourage just the top bud to develop. Insert the cutting so that the tip is barely covered or is at soil level.

SOFT-PITH AND HOLLOW-STEMMED CUTTINGS

Plants that have soft pith or hollow stems, such as forsythia, kerria and sambucus, may rot while the cuttings are rooting, and so will fail to root, unless any exposed hollow stem or soft pith is protected. Make the cuts at the nodes as normal and if the hollow stem or soft pith is exposed seal it by dipping the base of the cutting into melted candle (paraffin) wax. Make sure the wax is not too hot or it could damage the cutting. Alternatively, try taking a heel cutting (see p56).

The cuttings can generally be forgotten until the following year, as the bottom cut surface undergoes a period of callusing over the winter from which roots develop in spring. However, it is a good idea to check the trench after frosts and refirm the soil if needed. Ensure cuttings do not dry out during dry periods in late spring and summer.

Cuttings will have rooted within 12 months, when they can be lifted and planted in their permanent position.

• EVERGREEN CUTTINGS •

Evergreens are normally propagated from late summer to autumn from very ripe wood that is almost hardwood. They cannot be treated as hardwood cuttings as they still carry leaves and are not truly dormant because of their evergreen nature. Most evergreens can also be propagated from softer wood earlier in the year, and these are treated according to the condition of the wood – softwood, greenwood or semiripe.

The length of the cutting depends on the normal stem length for that plant; this could be as short as 5cm (2in) for dwarf hebes and up to 15cm (6in) for *Prunus laurocerasus* and *Magnolia grandiflora*. For plants that are difficult to root or those to be propagated in unsterilised soil in a cold frame, take heel cuttings (see p56) or mallet cuttings (see p57).

Trim just below a bud, if appropriate. Leave the apical (shoot-tip) bud intact, unless the top of the stem is still soft, in which case remove the soft tip and cut down the stem to just above a bud. On plants that have very large or awkward leaves, reduce the leaf area by cutting their leaves in half horizontally.

Prepare the soil in a cold frame by digging in plenty of organic matter and sharp sand to produce a good cuttings medium. Insert the prepared evergreen cutting so that the bottom leaves are in contact with the soil. Mist thoroughly with a fungicide and then seal the lid of the cold frame.

Alternatively, place evergreen cuttings in a pot filled with cuttings compost. Set it in a cold frame or other protected area where the aerial environment can be kept warm and moist; a large propagator will be fine, but do not use tight-fitting polythene bags or plastic covers.

Protect the cuttings from frost and cold weather. During sunny weather

SPECIAL CARE IN WINTER

A mist propagation unit (see p17) usually produces much better results with difficult-to-root subjects. However, you do need to take care during late autumn and winter if temperatures and light levels are too low, because these conditions may discourage the cuttings from developing roots, and the high humidity can encourage excessive leaf and stem rotting.

shade the propagator or cold frame with a shading wash, shade netting or fleece and keep the compost moist. Spray occasionally with a fungicide. As good rooting can take up to 12 months, keep a close eye on the cuttings. Remove fallen or diseased ones as soon as they are seen and generally care for the cuttings until they have rooted and are ready to lift and pot up individually.

PROPAGATING FROM EVERGREEN CUTTINGS

In winter prune the parent plant to encourage development of strong shoots, which will root more quickly.

In late summer take a heel cutting from the current growth. Trim the heel. Pinch out any soft tips.

Remove the leaves from the bottom third of the cutting, which should be 5–15cm (2–6in) long.

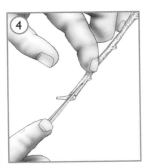

Remove a thin sliver of bark, 2.5cm (1in) long, from the bottom of any stem that may be difficult to root.

Treat the base with hormone rooting powder, then insert the cutting up to its bottom leaves. Water in.

In autumn transplant the rooted cuttings to their final growing position and water in well.

• HEEL & MALLET CUTTINGS •

Where standard stem cuttings prove difficult to root, slightly modified stem cuttings often have a much better success rate. These have a piece of older wood attached, which is less prone to rotting. The two methods are heel cuttings and mallet cuttings.

HEEL CUTTINGS

These can be taken from any type of stem – softwood, greenwood, semiripe and hardwood, including evergreens – and at any time of year. They are a useful way of propagating plants with pithy stems as well as autumn-propagated plants that would otherwise struggle through winter. Berberis, ceanothus and rosemary are commonly propagated from heel cuttings.

When removing a suitable sideshoot, ensure it comes away with a 'heel' or tail of bark from the main stem. If the sideshoot doesn't come away easily or cleanly, place a knife blade in the axil of the sideshoot and carefully cut away the sideshoot, ensuring you cut out a heel. Remove the leaves from the bottom third to half of the sideshoot and insert it in a pot filled with cuttings compost, until the bottom leaves are in contact with the soil. Place hardy cuttings into a cold frame and less hardy ones into a

PROPAGATING FROM HEEL CUTTINGS

Hold the bottom of a sideshoot between the thumb and forefinger. Pull down sharply.

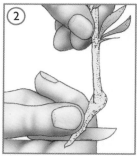
Neaten the long tail on the heel and any leaves near it. Dip the basal cut in rooting hormone.

Make a hole in the soil or compost. Plant the cutting. Label it and water with a fungicide.

more protected environment, such as a heated propagator. Hardwood cuttings can also be planted outside in well-prepared soil.

MALLET CUTTINGS

These cuttings are so called because they look like a hammer. They have several advantages over heel cuttings in that they contain more material for the new plant to root from and, having been removed from the plant with a neat cut, are less likely to die back or rot.

Mallet cuttings can be used only with semiripe and hardwood cuttings, such as those of mahonia, and are particularly useful for plants with hollow or pithy stems, such as berberis and spiraea, which are propagated in autumn and overwintered in a cold frame.

Take cuttings in late summer or autumn. Thick or very woody cuttings will root better if a sliver of bark is removed along the bottom of the mallet.

Insert the cutting into a pot of cuttings compost, so that the bottom leaves are in contact with the compost. Mist the cuttings with a fungicide.

Place semiripe and hardwood mallet cuttings into a cold frame and less hardy ones into a more protected environment, such as a heated propagator. Hardwood mallet cuttings can also be planted in well-prepared soil.

PROPAGATING FROM MALLET CUTTINGS

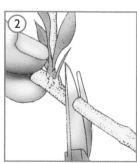
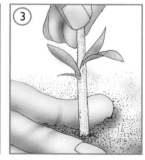

Cut horizontally with secateurs across the parent stem just above a suitable sideshoot.

Make a basal cut about 2cm (¾in) below the first cut. Remove any leaves from the bottom third.

Dip the basal cut in rooting hormone. Insert the cutting so the 'mallet' is buried, and label.

• LEAF-BUD CUTTINGS •

Leaf-bud cuttings are most commonly taken from semiripe stems although they can be struck from softwood, greenwood and evergreen shoots. Each cutting consists of a leaf (or part of a cut-down large leaf), a growth bud in the leaf axil and a small section of stem. The leaf supplies the food, the bud forms the new stem system and the stem produces the roots.

As there is only a short section of stem, and so little stored food reserves, the propagation material should have the optimum potential to produce roots. The selected piece of stem must have a healthy, fully developed leaf and a

DOUBLE LEAF-BUD CUTTINGS

With plants, like clematis, which have two opposite leaves/buds at the axil, prepare the cutting by cutting the stem just above the axil/bud. Then trim the stem 2.5–4cm (1–1½in) below the bud.

Then remove one leaf or, as with clematis, one leaflet from one or both leaves. Alternatively, split the stem vertically down the middle to make two cuttings, which are placed horizontally in the compost.

PROPAGATING FROM LEAF-BUD CUTTINGS

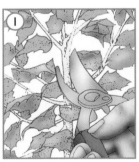

Prune the parent plant, if necessary, to encourage new stems with a high rooting potential.

Select a suitable stem with fully developed, mature leaves and healthy, viable buds in its axils.

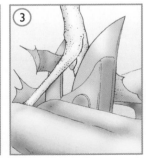

Make an angled cut on the selected stem as close above the bud/axil as possible to ensure no 'snag' remains.

healthy, viable bud in the leaf axil. An immature leaf may wilt or rot or continue to grow and use up the food reserves needed to produce new roots.

The short piece of stem below the bud/axil is used to anchor the cutting in the compost; the larger the leaf, the longer the stem that is needed.

To reduce the leaf area either cut an entire leaf in half horizontally or remove half the leaflets of more complicated leaves, such as mahonia.

Dip the bottom cut end of each cutting in rooting hormone, then insert it in a pot of cuttings compost. Mist the cuttings with fungicide. Place hardy leaf-bud cuttings into a cold frame and less hardy ones into a more protected environment, such as a heated propagator.

VINE-EYE CUTTINGS

This is the leaf-bud equivalent for hardwood cuttings, especially vines and other woody climbers. Cuttings are treated in the same way, but being hardwood they are leafless and easier to work with. To ensure good rooting it is advisable to wound the stem by removing a thin, vertical sliver of bark, 2.5cm (1in) long, on the side opposite the bud.

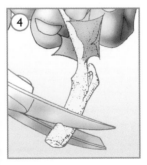

Cut the stem 2.5–4cm (1–1½in) below the bud/axil; exactly how far depends on the size of the leaves.

Reduce the size of the leaf if it is large to lessen water loss and make it easier to handle.

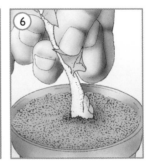

Insert in a pot of compost with its bud about level with the compost surface. Firm well and water in.

• INTRODUCTION TO LEAF CUTTINGS •

Of all the propagation methods, leaf cuttings are probably the most intriguing – a mature leaf produces new young plants (plantlets) on its surfaces. However, the range of plants that can be propagated in this way is limited.

The best leaves to use are those that have recently expanded to their mature size, as they will manufacture plenty of food and energy through photosynthesis, providing food for the new developing plantlets. As well as selecting suitable mature leaves you should ensure that they are typical of the plant, especially those with variegated foliage. In addition they should be whole, undamaged and free from pests and diseases.

MAKING THE CUTTINGS

Houseplants and indoor plants grown from leaf cuttings can be propagated all year round, whenever suitable leaves are available. However, they will produce plantlets quicker and more successfully in spring and summer.

Bulbous plants propagated by the monocot method (see p72) can be increased whenever good, healthy, strong foliage is available, and this will depend to some extent on when they flower.

As soon as the leaf is separated from the parent plant it will start to lose water and once too dry will never root. It is therefore important to prepare the leaves as soon as possible after removal and then place them in their propagating position promptly.

To promote successful plantlet production the leaves need very high humidity and warmth, so place them in a heated propagator with a close-fitting lid or cover the container with a polythene bag and set it somewhere warm and light but out of strong, burning sunlight.

WHOLE LEAVES

Many succulents, especially those grown for their attractive foliage, can be propagated from whole leaves.

Carefully remove a healthy, mature whole leaf from the base of the plant and leave it in a warm place (minimum 10°C/50°F) for 24–48 hours for the cut surface to dry out slightly and callus over before potting up. Insert the cut end into pots or seedtrays filled with two parts cactus compost to one part fine grit, topped off with fine grit, and firm in. Do not cover. Place in a warm position (21°C/70°F) and good light. Keep the compost just slightly moist.

SUITABLE PLANTS FOR LEAF CUTTINGS

Many of the plants that are suitable for leaf cuttings are popular houseplants, so you can use this easy and efficient method of propagation to increase and rejuvenate the plants in your home.

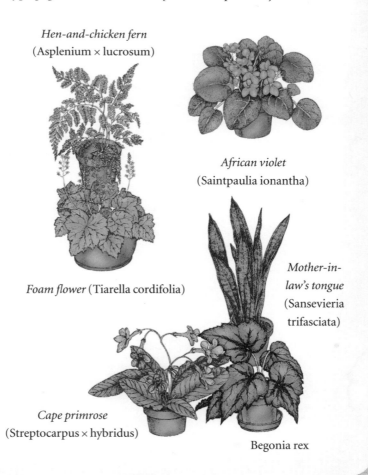

Hen-and-chicken fern (Asplenium × lucrosum)

African violet (Saintpaulia ionantha)

Foam flower (Tiarella cordifolia)

Mother-in-law's tongue (Sansevieria trifasciata)

Cape primrose (Streptocarpus × hybridus)

Begonia rex

• LEAF-STALK CUTTINGS •

Sometimes known as leaf-petiole cuttings, this is one of the simplest and most reliable ways to produce new plants from leaf cuttings. It is especially suitable for peperomia and African violets (*Saintpaulia*).

Cuttings can be taken at any time of year, providing a new but fully expanded/matured leaf is available. Cuttings will root more quickly and easily if taken in spring and early to midsummer. Always select perfectly healthy leaves and avoid any that are even slightly damaged or diseased.

Leaf-stalk cuttings are very susceptible to rotting and disease, so ensure that knife blades and containers are scrupulously clean and that fresh compost is used. This can be a standard cuttings compost or a mixture of equal parts by volume multipurpose compost and sharp sand or perlite. Use seedtrays where a lot of plants are needed, or shallow pan pots for a few plants.

Angle the leaf stalk in the compost so that the blade is almost flat on the surface; the shallower the angle it is inserted, the more quickly the cutting will root. Providing the compost is moist, there is no need to water in.

Place the container in a warm, humid environment – essential for

PROPAGATING FROM LEAF-STALK CUTTINGS

Remove an undamaged leaf plus leaf stalk, and trim the leaf stalk to 4–5cm (1½–2in) long.

Make a shallow hole and insert the cutting as shown, with the leaf blade in contact with the compost.

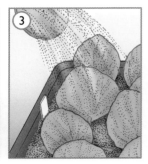

Firm the compost gently around the base of the cutting. Label and apply a fungicide.

good rooting. Preferably use a heated propagator, providing a steady bottom heat of around 21°C (70°F), or a sealed clear polythene bag and put in a warm place. Ensure good light is available but keep out of strong, direct sunlight.

Cuttings are prone to rotting off if the compost is too wet. Water carefully and air the cuttings once or twice a week. Remove any decaying leaves promptly.

New plantlets will form on the cut surface of the leaf stalk, usually within 5–6 weeks, and more than one plantlet may develop at each point. When plantlets form, remove the cover and allow the plantlets to grow on until large enough to handle and pot up individually. Plantlets that take a while to reach a size large enough to handle will need feeding with a liquid fertiliser in the interim.

Place the cuttings in light shade in a propagator that is heated from below.

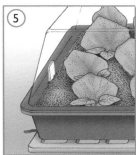

Grow on until large enough to handle. Apply a liquid feed to any plantlets that lag behind.

Pot up the plantlets individually. Label, water in and then harden them off.

• MIDRIB CUTTINGS •

Some plants, such as *Sinningia* (*Gloxinia*) and *Streptocarpus*, have a large strengthened vein along the midline of their leaves and can be propagated from these leaves. Midrib cuttings can be taken at any time of year, providing that the plant has a recently fully expanded, undamaged leaf. (Old leaves are less likely to root quickly.) Cuttings will root more quickly and easily if they are taken from midspring to midsummer.

Midrib cuttings are very susceptible to rotting and disease, so clean knife blades and containers are essential. Use seedtrays or shallow pan pots depending on the number of plants you want.

Fill the container with fresh standard cuttings compost or mix together equal parts of multipurpose compost and sharp sand or perlite. Lightly firm the compost so that it settles to within 1cm (½in) below the rim. Ensure that the compost is moist, to avoid having

New young plants developing at the base of a Cape primrose midrib cutting.

to water once the midrib cuttings are inserted into it.

When slicing the leaf into strips use a sharp blade so you do not rip the leaf. Leave the cuttings on the cutting surface to make it easier to remember which is the bottom of the cutting (nearest the bottom of the leaf).

Insert the midrib cuttings, bottom end down, 2.5–4cm (1–1½in) apart and spray them with a fungicide to protect against disease and rots. Even so, the cuttings are prone to rotting off if the compost is too wet. Always water carefully from below by standing the tray or pot in a shallow container of tepid water. Allow the compost to become wet through, but do not let it become saturated or waterlogged. Remove the container and allow to drain before returning it to its propagation area. Air the cuttings once or twice a week and remove any decaying leaves promptly.

New plantlets will form on the lower cut surface of the leaf at the midrib, usually within 5–8 weeks. When plantlets form, remove the cover and allow the plantlets to grow on until large enough to handle and pot up individually. The compost may need rewetting during the rooting process, but never water from above, only from below. Meanwhile feed the plantlets with a liquid fertiliser.

PROPAGATING FROM MIDRIB CUTTINGS

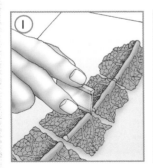

Remove an undamaged leaf, turn upside down and cut into strips no wider than 5cm (2in).

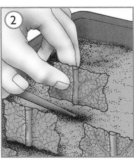

Make a shallow trench and insert the cuttings deep enough to stay upright. Gently firm in.

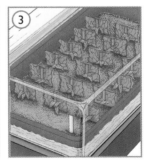

Place the container in a heated propagator. If necessary, rewet drying compost from below.

• LATERAL VEIN CUTTINGS •

This is an alternative way of increasing plants that have a single, large, central leaf midrib, such as gesneriads, sinningia and Cape primrose (*Streptocarpus*).

Lateral vein cuttings can be taken at any time of year, providing the plant has an undamaged leaf that has only recently fully expanded. They will root more quickly and easily if taken from midspring to midsummer. Such cuttings are very susceptible to rotting and disease, so clean containers are essential.

Use fresh standard cuttings compost or mix equal parts by volume of multipurpose compost and sharp sand or perlite. Fill a seedtray and lightly firm the compost so that it is within 1cm (½in) of the rim. Leave it to soak in a tray of tepid water until the compost is moist, to avoid having to water once the cuttings are inserted.

Carefully remove a suitable complete leaf from the parent plant using a sharp, clean knife or blade. Ensure you remove it all, as leaving a small section behind can induce rotting in the parent plant.

When cutting the leaf close to the midrib make sure that the midrib is completely removed so that all the lateral veins have an exposed cut surface.

Insert each leaf cutting vertically, with the cut surface just in the compost but

PROPAGATING FROM LATERAL VEIN CUTTINGS

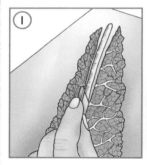

Lie a leaf upside down on a clean sheet of glass. Remove the midrib with a sharp knife or razor blade.

Make a shallow trench in moist cuttings compost. Insert each cutting with the cut surface facing down.

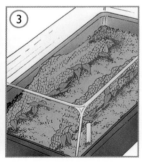

Label clearly. Place the seedtray in a propagator. Pot on plantlets once they can be handled.

Gorgeous Cape primrose is easy to propagate from leaf cuttings.

deep enough to keep it upright. Gently firm in each cutting. Then spray the lateral vein cuttings with a fungicide to protect against disease and rots.

Place the container in a warm, humid environment, which is essential for good, fast rooting. Preferably use a heated propagator, providing a steady bottom heat of around 18–21°C (65–70°F), or else cover the seedtray or pot with a clear polythene bag and leave it in a warm place. Ensure good light is available to the cuttings but keep them out of strong, direct sunlight.

Air the cuttings once or twice a week and remove any decaying leaves promptly to avoid infection.

Lateral vein cuttings are prone to rotting off if the compost is too wet or if watered from above, so water carefully when the compost needs rewetting during the rooting process. Always water from below by standing the seedtray or pot in a shallow container of tepid water. Allow the compost to become wet through, but do not let it become saturated or waterlogged. Remove the container and allow to drain before returning it to its propagation area.

New plantlets will form on the cut surfaces of each lateral vein within 5–8 weeks. When plantlets form, remove the cover and allow the plantlets to grow on until they are large enough to handle and pot up individually. As plantlets take a while to be large enough to handle, they will meanwhile need feeding with a liquid fertiliser.

• LEAF SLASHING •

Plants that do not possess a central midrib and lateral veins, but instead have a more netted veining system, such as *Begonia rex* and related species and *sinningia*, can be propagated by cutting through the veins – that is, by leaf slashing. The leaf itself is kept intact and plantlets form at each cut.

Cuttings can be taken at any time of year, providing the plant has a recently fully expanded, undamaged leaf. Avoid any that are diseased or even slightly damaged, and old leaves which are less likely to root quickly. Cuttings will root more quickly and easily if taken from midspring to midsummer.

Clean all tools and equipment thoroughly as leaf cuttings are very susceptible to rotting and disease.

Fill a seedtray with standard cuttings compost or mix equal parts of multipurpose compost and sharp sand or perlite. Lightly firm the compost so that it settles to within 1cm (½in) of the rim. Moisten the compost thoroughly.

Carefully remove a suitable leaf from the parent plant using a sharp knife or blade. Ensure you remove the complete leaf stalk, as leaving a small section behind can induce rotting.

Place the leaf upside down on a pane of glass or cutting board. Cut off the leaf

PROPAGATING BY LEAF SLASHING

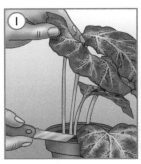

Remove a fully expanded, undamaged leaf with its leaf stalk and cut off the stalk.

Place the leaf upside down and make cuts 6–13mm (¼–½in) long across the major veins.

Place the leaf top side up on the compost surface. Pin down with wire staples and small pebbles.

CUTTINGS

stalk flush to the leaf blade. Make a series of 6–13mm (¼–½in) long cuts across major or large veins at right angles to the vein, roughly 2.5cm (1in) apart. Having secured the leaf on the compost, spray it with a fungicide to reduce the likelihood of disease and rots.

To maintain high humidity cover the seedtray with a sheet of glass or put it in a clear polythene bag. Set the seedtray in a warm, humid environment, such as a heated propagator providing a steady bottom heat of 18–21°C (65–70°F). Place in good light, which is essential for food production and so for plantlet development. Avoid strong, direct sunlight, which may scorch the leaf.

New plantlets will form at each cut within 3–5 weeks. In the meantime little or no further watering should be needed if the compost was well watered initially and the seedtray well covered. If the compost dries out, stand the tray in a shallow container of tepid water. Allow the compost to become wet through, but do not let it become saturated or waterlogged. Remove the container and allow to drain before returning it to its propagation area.

When plantlets are well formed, remove the cover and allow the plantlets to grow on until large enough to pot up individually; the original leaf will usually have started to rot away by then.

Cover the tray with a sheet of clean glass. Place in a warm, shaded environment.

Leave the developing plantlets in place until they have developed a good root system.

Once they are large enough to be handled, separate the plantlets and pot on individually. Harden off.

• LEAF SQUARES •

Any leaf that will regenerate vegetatively can be propagated from leaf squares but this method is normally used for plants with big leaves and especially for *Begonia rex* and related species and sinningia (*Gloxinia*). Its main advantage is that it produces numerous plantlets from a single leaf.

PROPAGATING FROM LEAF SQUARES

Fill a container with compost and firm to within 1cm (½in) of the rim. Water well and then drain.

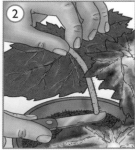

Carefully remove a healthy, fully expanded, undamaged leaf from the plant.

Place the leaf face down on a sheet of glass. Cut into 2–2.5cm (¾–1in) squares using a blade and ruler.

Lay the leaf squares face up on the compost 6–13mm (¼–½in) apart and spray with a fungicide.

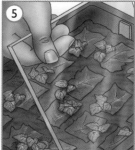

Cover with a pane of glass. Place in a warm, shaded area. Harden off once the plantlets produce leaves.

When the plantlets are large enough to handle, pot up individually. Water in and label.

Propagation by leaf squares can be done at any time of year, providing the plant has a recently fully expanded, undamaged leaf. The cuttings will root more quickly and easily if taken from midspring to midsummer. Old leaves are less likely to root quickly.

Small pieces of cut leaf are very prone to rotting and disease. Therefore clean, sterile tools and equipment should be used at all times and scrupulous, hygienic measures observed. Use fresh standard cuttings compost or mix equal parts of multipurpose compost and sharp sand or perlite.

Fill a seedtray with compost and lightly firm so that the compost is within 1cm (½in) of the rim. Water it well.

Remove a suitable leaf from the parent plant using a sharp knife or blade. Always select a healthy leaf and avoid any that are even slightly damaged or diseased. Ensure you remove the complete leaf stalk, because leaving even a small section behind can induce rotting in the parent plant.

When slicing the leaf into squares, make sure that each square has a large vein. Discard any squares that are damaged. Having set the cuttings flat on the compost, spray them with a fungicide to protect against disease and rots.

> **WRINKLED OR HAIRY LEAVES**
>
> Wrinkled or very hairy leaves should be inserted vertically in the compost, just deep enough to keep the squares upright. Ensure that the basal cut end — the cut nearest the base of the leaf — is in the compost.

Cover the seedtray with a clean sheet of glass or put in a clear polythene bag to maintain high humidity. Place the tray in a warm area, preferably a heated propagator providing a bottom heat of 18–21°C (65–70°F). Ensure good light is available but keep out of strong, direct sunlight. A windowsill in a warm room facing east or west might be suitable.

If the compost dries out, stand the tray in a shallow container of tepid water until the compost has become wet through. Drain the container before returning it to the propagation area.

New plantlets will form on the cut surfaces of the larger veins within 5–6 weeks. When plantlets form, remove the cover and allow the plantlets to grow on until large enough to pot up individually, which normally takes several more weeks.

• MONOCOT LEAF CUTTINGS •

Monocotyledonous plants (that is, flowering plants producing embryos with a single cotyledon/seed leaf), such as snowdrops (*Galanthus*), Cape cowslips (*Lachenalia*), snowflakes (*Leucojum*), pineapple lily (*Eucomis*) and mother-in-law's tongue (*Sansevieria*), have a series of parallel veins running the length of the leaf. Many of these plants can produce a new plantlet at the cut surface of the leaf veins.

Some of these leaves, especially those of bulbous plants, tend to wilt very quickly, so keep them turgid and prepare them as soon as possible after collection. Monocot leaf cuttings are also very soft and prone to rotting, so ensure you propagate from them in hygienic

PROPAGATING FROM MONOCOT LEAF CUTTINGS

Fill a container with cuttings compost. Water well and allow to drain. Cut off a fully expanded, undamaged leaf. Lie it down on a sheet of glass.

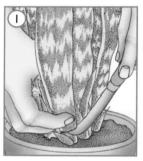

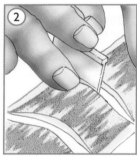

Cut the leaf into a series of sections 2.5–4cm (1–1½in) wide by slicing at right angles to the veins with a sharp knife or razor blade.

Insert the cuttings vertically in shallow trenches 2cm (¾in) deep and 2.5cm (1in) apart. Lightly firm in, label the seedtray and then spray with a fungicide.

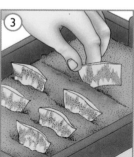

Place the container in a warm, humid environment. Rewet the compost if it dries out by watering from below.

conditions with extremely clean tools and equipment. It is also essential to provide conditions that will allow the leaves to root quickly.

To help reduce the likelihood of drying out fill the container with compost before taking the cuttings from the parent plant. Use a standard cuttings compost or mix equal parts by volume of multipurpose compost and sharp sand or perlite. Ensure that the compost is moist, to avoid having to water once the cuttings are inserted into it.

Cut the leaf into a series of wide sections or cut it into shallow chevrons. Make a shallow trench in the compost with a dibber, then insert the lower edge of each monocot leaf cutting just deep enough to keep it upright. Make sure you keep each cutting facing the same direction as it was growing previously on the parent plant.

Place the container in a heated propagator, providing a bottom heat of 18–21°C (65–70°F), or in a clear polythene bag, supported above the cuttings with hoops of metal, and keep in a warm place. Ensure the cuttings receive plenty of good light so they produce food, although strong, direct sunlight should be avoided otherwise scorching may occur.

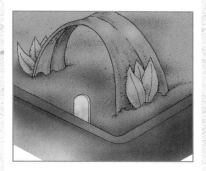

HOOPING LEAF CUTTINGS
Leaf cuttings of helionopsis will regenerate from both ends. Reduce a leaf to 4–5cm (1½–2in) by cutting off the top and bottom ends. Plant the cutting in a hoop, with both ends set into the compost.

If the compost begins to dry out, water carefully from below, then drain well as cuttings are prone to rotting off if the compost is too wet. Air the cuttings once or twice a week and remove any decaying leaves promptly.

New plantlets will form on the lower cut surface of each monocot leaf cutting, usually within 5–8 weeks. When this occurs, remove the cover and allow the plantlets to grow on until large enough to handle and pot up individually. Then label each pot and harden off the new plantlets (see p16).

• INTRODUCTION TO ROOT CUTTINGS •

Root cuttings are taken when the plants are fully dormant, in late autumn or winter. They can be easier to care for than some stem cuttings and they are less likely to dry out or rot. However, they can be relatively slow to develop, especially on old herbaceous plants. To increase your chances of success, you should first prepare the parent plant (see p76) before using it for propagation.

Each parent plant produces great numbers of new plants from root cuttings, and these new plants are relatively large and vigorous and need very little in the way of aftercare. Another bonus is that plants from root cuttings are free of foliar pests and diseases, such as stem and leaf nematodes, which may affect their parents. Root cuttings, however, cannot be used to propagate variegated plants, as the new plants will have plain leaves.

Herbaceous plants that take particularly well from root cuttings are those with thick or fleshy roots. These include *Acanthus*, *Anemone hupehensis*, *A. × hybrida*, *Campanula*, globe thistle (*Echinops*), sea holly (*Eryngium*), *Papaver orientale*, *Phlox*, *Primula denticulata*, pasque flower (*Pulsatilla vulgaris*), tree poppy (*Romneya coulteri*) and *Verbascum*.

Also a few woody plants can be propagated in this way, such as *Aesculus parviflora*, tree of heaven (*Ailanthus*), *Aralia*, *Catalpa*, *Chaenomeles*, *Clerodendrum*, *Robinia*, *Sophora* and *Syringa*, and climbers such as *Campsis*, passionflower (*Passiflora*) and *Solanum*.

Sea holly is easy to propagate from root cuttings.

• WOODY SHRUBS WITH SUCKERS •

Naturally suckering trees, shrubs and climbers as well as raspberries, for example, are propagated by severing the rooted suckers – shoots arising from the roots – from the parent plant and replanting them once they have had a chance to become established.

Suckers develop quickly as new plants. Because they already have their own root system, they can readily be transplanted where you want to grow them. The drawback is that you have to wait for suckers to appear naturally, although you can sometimes encourage them to form by wounding the main root. However, this is not necessarily a good idea, as wounding may not produce suckers and it may cause disease to enter the root. Also, once done, wounding can encourage a thicket of shoots to develop.

Grafted plants that are grown on a rootstock, including trees, shrubs, roses and fruit trees, should never be propagated from suckers. Because these arise from the rootstock, instead completely remove them at their base as soon as they are seen. It is best to carefully pull them off, rather than cut them off. If this is not done, the grafted plant will become weakened and taken over by a thicket of young rootstock suckers.

PROPAGATING WOODY SHRUBS FROM SUCKERS

During the dormant season carefully lift isolated suckers from woody shrubs.

Trim back roots, especially damaged ones, and some of the topgrowth to reduce stress on the roots.

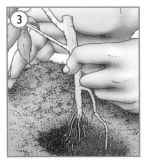

Replant each new plant in well-prepared soil at the same depth that it was previously growing.

• PREPARING THE PARENT PLANT •

Although it is possible to take root cuttings from mature herbaceous plants as they are growing in the ground, it can be a stressful experience for the parent plant, especially if the plant is too young or very old.

If the plants are too young, with little suitable root growth, it is worth waiting

PREPARING THE PARENT PLANT FOR ROOT CUTTINGS MATERIAL

Carefully lift a healthy plant from the ground during the dormant season.

Cut back most of the topgrowth. Shake off any excess soil from the roots.

Wash the roots in a bucket of water or hose them clean to remove the rest of the soil.

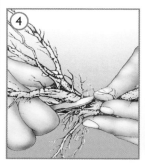

Cut off the very fine roots using a sharp knife; leave the average-sized roots.

Replant the rootstock in well-prepared soil at the same level as it was previously growing.

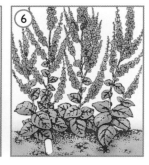

Allow the plant to reestablish during the growing season; it will produce vigorous, new roots.

a year or trying to induce further root growth before attempting to remove cuttings material.

The larger roots of old or tired plants will not have good potential to produce stem buds and so generate new plants. Some plants may simply not have produced roots of sufficient size and quality to propagate from.

In these situations, and where you have the time and the opportunity, it is a good idea to prepare the parent plant a year in advance before using it to supply propagation material. The new, vigorous roots that will develop through the year will produce stem buds and new plants far more readily and increase your chance of success.

When the parent plant is fully dormant – this is usually from midautumn to midwinter – lift it with all its roots. Cut back the topgrowth to as near the crown as possible, preferably within 5–10cm (2–4in), especially removing old growth and dead and diseased leaves and stems. Very large clumps may need to be split into smaller, more manageable sections and the oldest, middle sections disposed of.

Reduce the root system by cutting back the largest, major roots to within 5–10cm (2–4in) of the crown, with a sharp knife. Very fine, dead or damaged roots should also be removed, to leave average-sized roots, which may also need dead or damaged areas cut off.

Replant in well-drained soil to which has been added plenty of organic matter and even sharp sand if necessary. Alternatively, replant the parent plant in a large pot of good-quality compost.

At this point, it may be worth propagating from some of the largest roots you have cut off while preparing the parent plant for future cuttings material. If they do root you will have created new plants a year earlier.

Look after the parent plant during the growing season, watering and feeding as necessary, to ensure it produces plenty of new root growth. Remove flowering growth to prevent energy being wasted on the production of flowers and seeds.

The roots that develop at the beginning of the growing season will grow the most quickly through the year and so have the best potential for producing new plants. Therefore when the time comes to lift the parent plant again to obtain the cuttings material, it is best to take your cuttings from its largest roots and from the top of the root closest to its crown.

• OBTAINING MATERIAL & PREPARING A CUTTING •

Whereas some herbaceous plants will produce new plants from root cuttings at any time of year – horseradish, for example – nearly all other plants should be propagated from root cuttings in the middle of their dormant period. For most plants this means winter.

If you want only one or two cuttings, scrape away the soil from around the crown of the root to expose the roots. Cut off a suitable root close to the crown. This method is also suitable for plants, such as sea holly (*Eryngium*) and pulsatilla, that resent root disturbance.

Where more cuttings are required or suitable roots cannot be found by scraping away the soil, you will need to lift the plant to obtain the cuttings material, which should be of more-or-less uniform thickness. Replant the parent plant so that it can go on growing.

For root cuttings of woody plants, carefully dig down in the soil to expose part of the root system. Select roots up to finger thickness. Remove any fibrous roots and cut into 5–15cm (2–6in) sections. Then treat in the same way as root cuttings of herbaceous plants.

OBTAINING CUTTINGS MATERIAL

In the middle of its dormant season cut off any topgrowth and carefully lift the plant.

Wash the soil from the plant's roots. Then cut off the young roots close to the crown and set aside.

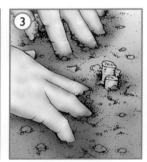

Return the parent plant to its usual position in the garden to continue growing on.

OPTIMUM SIZE FOR A ROOT CUTTING

As with any cutting, success will depend on the size of the propagating material and so the amount of food reserves stored in it; the bigger the cutting, the more food is stored and so the longer it can survive before rooting and producing leaves and shoots. Success also depends on temperature: the cutting develops leaves and stems more slowly in lower temperatures, so it needs more stored food than one rooted in warm conditions.

Cuttings rooted outdoors need to survive for 16–20 weeks before establishing and so should be 10–15cm (4–6in) long. In a cold frame or cold glasshouse, rooting should take place in about eight weeks and so cuttings need to be 5–7.5cm (2–3in) long, while in a warm glasshouse or heated propagator at a temperature of 18–24°C (65–75°F) they will take only about four weeks to root, so these root cuttings can be 2.5–4cm (1–1½in) long.

PREPARING A ROOT CUTTING

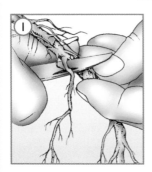
On roots that are firm and plump cut off any fibrous, lateral or side roots close to the main root.

Make a straight cut where the root joined the parent plant so you know which way the root was growing.

Trim the cutting to the correct length by cutting away the thin, basal end, using a sloping cut.

• STARTING ROOT CUTTINGS •

Although root cuttings can be inserted into well-prepared soil in a cold frame, it may be more convenient to start them off in pots. The size of pot depends on the number of cuttings you have. If they are spaced 2.5–4cm (1–1½in) apart, you can plant seven root cuttings in a 9cm (3½in) pot. Fill it with multipurpose compost or seed and cuttings compost.

Hormone rooting compounds used to induce roots on stems should not be used on root cuttings because they will actually inhibit bud formation.

The best way to protect a root cutting against rots and disease is to dust it with a fungicidal powder.

After inserting all the treated cuttings, lightly tap the pot to settle

STARTING ROOT CUTTINGS

Fill a pot with compost. Make a hole with a dibber and insert each cutting vertically, with its top just visible.

Space the cuttings 2.5–4cm (1–1½in) apart. Then cover the compost surface with horticultural grit or vermiculite.

Strike off the grit/vermiculite until level with the rim. Label, and place in a cold frame or propagator in a shaded position.

Water only when roots have appeared through the drainage holes in the pot. Then apply a liquid feed and move to a light position.

ROOT DISTURBANCE

Some plants, such as tree poppy (*Romneya coulteri*), do not like to have their rooted cuttings lifted and disturbed. Therefore place just one or two cuttings in a small pot and treat as one plant. Transplant the rooted plants to their growing position, handling only the rootball carefully and avoiding the roots.

the compost around them. A covering of horticultural grit (or vermiculite for quick-rooting plants) will prevent the tops of the cuttings and the latent buds drying out. The weight of grit tends to compress the compost slightly, so causing the tops of the root cuttings to be pushed further into the grit. This provides almost perfect aeration for bud development and reduces the likelihood of rotting because the grit/vermiculite is less moist than compost. Such a covering also helps prevent algal growth. Do not water in the cuttings.

Place the pot in a propagating environment that is appropriate to the size of the root cuttings (see box p79).

Often new shoots emerge from the cuttings before they have produced new roots. Start watering then but only minimally until the plants are well established and have developed a new root system; look for roots emerging through drainage holes in the pot.

Once they are rooted and growing well, give a light liquid feed and move the pot to a well-lit position. Gradually increase watering. Harden off any young plants propagated in a heated propagator before moving permanently to cooler conditions (see p16). Plant out other rooted cuttings individually in a cold frame or grow them on in pots of potting compost.

THIN ROOTS

The roots of some herbaceous perennials, such as Japanese anemone, campanula and phlox, are thinner and therefore need to be treated differently. Choose root sections 8–13cm (3–5in) long. Place these horizontally, 2.5cm (1in) apart, on the compost, cover with 1cm (½in) of compost and topdress with grit.

TUBEROUS ROOT CUTTINGS

Some plants store food in swollen areas on their roots, known as tuberous roots – not to be confused with true tubers, which are swollen underground stems (see p92). They include alstroemeria, begonia, dahlia and foxtail lily (*Eremurus*). There are two types of tuberous roots – annual and perennial (see box right).

Many of the plants suitable for growing from tuberous roots are not hardy, so the roots need to be lifted in autumn, cleaned and stored over winter in a frost-free place below 5°C (41°F).

Although tuberous roots can be propagated by division, it is usually better to start them into growth in pots or trays with warmth and then take stem, leaf or leaf-bud cuttings from the resulting growth. Such plants usually have more vigour and perform better than those produced from divisions.

At the beginning of the growing season, divide a tuberous root so that each section contains at least one healthy growth bud. On annual tuberous roots, the buds will be in the crown at the base of the old shoots; the tuberous root does not have

Dahlia (annual roots)   *Begonia* (perennial roots)

TYPES OF TUBEROUS ROOTS

Annual tuberous roots, from plants such as dahlia, are produced every year around the crown or main stem. The energy stored in the tuberous roots is used during the subsequent year to produce new growth. As this happens the old tuberous roots shrivel and die.

Plants such as begonia produce just one large perennial tuberous root. This develops the shoots and fibrous feeding roots during the growing season. The tuberous root persists from year to year and enlarges sideways.

any buds itself and cannot reproduce itself. On perennial tuberous roots, there will be a cluster of buds in the centre of the crown.

Leave the divisions in a dry, airy place at 18–21°C (65–70°F). Once the cut surfaces seal themselves and produce a corky protective layer, pot up the divisions in moist, good-quality potting compost and place in a cool but frost-free place. There is no need to water. When new shoots appear move the pots to a warm, well-lit place so that strong, stocky growth is encouraged. You can also start to water the plants carefully, ensuring the compost is kept moist but not too wet.

Once good-sized plants have developed harden them off (see p16) before planting outside once there is little risk of frost.

DIVIDING ANNUAL TUBEROUS ROOTS

At the end of the growing season lift a plant. Clean the crowns thoroughly and dust with sulphur powder.

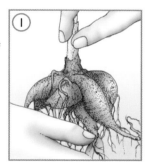

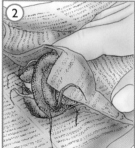

Wrap up the plant in newspaper. Store in a frost-free place until the buds begin to swell.

At the start of the next growing season divide the swollen roots into portions, each with at least one crown bud. Dust all cut surfaces with sulphur.

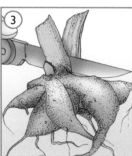

Leave in a warm, dry, airy place for about two days. Once the cut surfaces have formed a corky layer, pot up the cuttings.

CUTTINGS

• INTRODUCTION TO BULBOUS PLANTS •

This section covers plants that are either true bulbs or are described as being 'bulbous' in nature, which includes corms, tubers and rhizomes. All these structures are used to store food reserves and nutrients to keep the plant alive during dormancy and unfavourable growth periods, such as drought. They are also the way in which these plants spread and produce new plants. Such storage organs are known as modified stems, and so must have all the characteristics of a stem (see opposite and p39): an apical growing point, the stem itself with leaves with buds in their axils and flowers. The stem usually grows at or below soil level and may even grow horizontally.

BULBOUS PLANT TYPES

TUBER
- Node with lateral buds
- Internode
- Apical bud
- Root

CORM
- Apical shoot
- Internode
- Node
- Scale leaf
- Root

RHIZOME
- Node with lateral bud
- Internode
- Apical bud
- Leaf
- Root

BULB
- Flower bud
- Foliage leaf
- Scale leaf
- Lateral bud at leaf axil
- Root
- Stem

A true bulb has a short, fleshy, usually vertical stem, which, at its apex, has a growing point enclosed by thick, fleshy scale leaves, which are attached to a basal plate at their bottom end. The roots are produced from the basal plate. The foliage and flower buds grow up from the centre of the bulb.

A corm is a swollen base of a solid stem that stores the food reserves and which is surrounded by scale leaves. These are attached to the stem at distinct nodes, which have a bud in each axil. At the apex of the corm is a bud that will develop into the leaves and the flowering shoot. One of the major differences between bulbs and corms is that corms last just one year whereas most bulbs can last for several years.

Tubers have 'eyes', which have a cluster of buds and a leaf scar. They are arranged around the apical bud on the end opposite the scar where the tuber was attached to the parent plant.

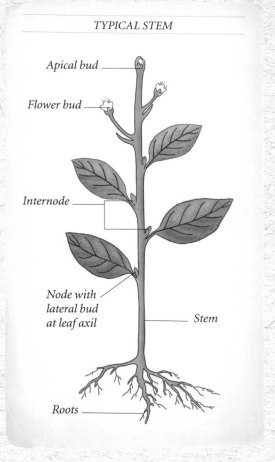

Rhizomes grow horizontally on or just below the soil surface. The stem appears segmented as it is made up of nodes and internodes, the leaves and flower buds developing from apical or lateral buds on the stem. Roots and a lateral shoot also grow from these points.

• BULBS •

True bulbs are among the easiest plants to propagate. As well as by other more specific methods, most bulbs readily multiply by naturally producing offsets (bulblets) from the parent bulb. Bulblets can be removed when bulbs are lifted

PROPAGATING FROM BULBLETS

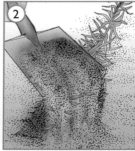
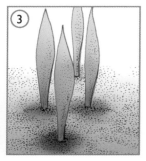

Pinch out any buds or flowers, then twist the stem out of the bulb, leaving the bulb in the ground.

Lay the bottom two thirds of the stem in a sloping 15cm (6in) deep trench. Fill the trench with sand.

In autumn dig up the stem. Remove the resulting bulblets and replant at twice their depth.

PROPAGATING FROM BULBILS

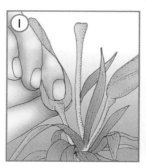
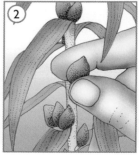

Remove all the flower buds from a suitable lily species just before they start to flower.

Bulbils will develop in the leaf axils and can be carefully picked off once dark and mature.

Press them into the compost in a pot, spacing them 2.5cm (1in) apart. Then cover with grit.

for storage or when too many bulblets reduce overall flowering. They will be identical to the parent bulb, making them suitable for propagating cultivars as well as species.

Detach the bulblets and, if necessary, dust with sulphur to prevent fungal rots. Larger, hardy bulblets can be replanted in the ground immediately, while small or tender bulblets benefit from growing

PROPAGATING FROM BULB SCALES

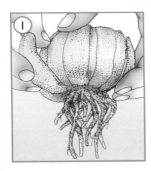

Remove a healthy scale leaf from the bulb by pressing it outwards or cutting it close to the basal plate.

Dust the scale leaves with sulphur. Place in a bag with a 50:50 mix of damp peat substitute and perlite.

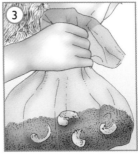

Fill the bag with air. Then seal and label it. Leave in a warm, dark place (21°C/70°F) for six weeks.

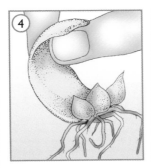

As soon as bulblets appear on the broken basal surfaces, remove the scale leaves from the polythene bag.

Plant up each scale leaf with its tip just visible. Harden off after the bulblet has produced leaves.

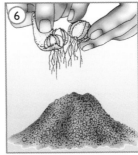

Once their leaves have died down, separate new bulblets from their scale leaf. Replant and grow on.

on in containers until they have reached a larger size. Smaller bulbs may take 2–4 years to flower, but some (such as *Cardiocrinum giganteum*) may need five to seven years.

Some lilies produce bulblets while others develop bulbils, which can be found in the leaf axils of species such as *Lilium bulbiferum*, *L. sargentiae* and *L. tigrinum* in late summer or autumn.

When ripe these detach easily and be potted up (see p86). Keep frost free over winter, and plant out the young plants the following autumn.

To increase a particular variety of bulb rapidly it is necessary to use artifical techniques such as propagating from bulb scales (see p87). Lift and clean a mature, virus-free bulb in late summer or early autumn. Discard any damaged

PROPAGATING BY SCOOPING BULBS

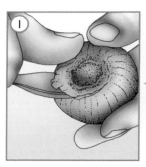

Carefully remove the basal plate, ensuring the base of every scale leaf is included. Then dust the cut scale leaf surface of the bulb with sulphur.

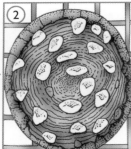

Place the bulb upside down on a wire rack or a seedtray filled with damp sand. Put it in a dark place at 21°C (70°F) until bulblets appear on the cut surface.

Plant the bulb upside down in a small pot of compost with the bulblets just below the compost surface.

At the end of the season, lift the remains of the old bulb and separate the bulblets. Replant at once.

outer scales. Having snapped off a few scales from the bulb as close as possible to the base, put these in a dark place at 21°C (70°F) for six weeks. Some lilies, such as *L. martagon*, need a further six weeks at 5°C (41°F).

When bulblets appear at the base of the scales, pot on individually, covered with their own depth of compost. If the scales have gone soft, remove them from the bulblets before potting on. If the scales are still firm, or have roots coming from their base, leave them attached to the bulblets.

Shallow planting stock bulbs or notching the basal plate can also promote the formation of bulblets.

SCOOPING AND SCORING

Scooping and scoring are forms of propagation used for bulbs such as hyacinths. In scooping, the centre of the basal plate is scooped out and discarded so that just the outer edge of the basal plate remains intact, while scoring involves making two shallow cuts at right angles into the basal plate. These shallow cuts or scoops cause callus tissue to develop, which stimulates the growth of bulblets. The treated bulb is placed in the dark while bulblets develop. Alternatively, after scoring, simply plant the parent bulb in compost with the cut surface downwards and wait for bulblets to form.

PROPAGATING BY SCORING BULBS

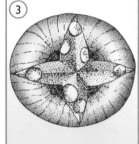

Make two cuts at right angles to each other across the basal plate of the bulb.

Put in a warm, dry place for a day or so until the bulb case opens out. Dust with sulphur.

Place on a tray and put in an airing cupboard until the bulblets develop. Then treat as scooped bulbs.

• CORMS •

Corms are similar to bulbs, storing food reserves to keep the plant going from year to year, but they have evolved from different structures. Corms are modified underground stems made from solid tissue and usually flattened in shape.

At the start of the growing season, new foliage and then flowers develop from the corm. When the foliage dies down at the end of the growing season, new corms develop at the base of each shoot on top of the old withered corm. Increase therefore is directly related to the number of stems produced by a corm. Normally most plants developing corms will propagate naturally to give sufficient new plants. However, if it is necessary to bulk up supplies more quickly, an artificial technique should be used. Always purchase corms from a reputable specialised grower, because it is vital to propagate from disease-free corms.

Some plants, such as montbretia (*Crocosmia*), produce their corms in long 'chains'. This means that each corm naturally divides into several over a couple of years to produce a thick clump, which may need splitting every few years to improve flowering.

To remove the corms without damaging them, carefully dig down

DIVIDING CORMS

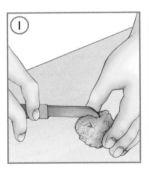

Just before planting in autumn cut a corm into several pieces, each with at least one bud.

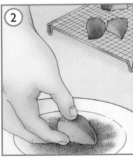

Dust the cut surfaces with sulphur. Set aside on a wire rack and leave in a warm, dry place.

Once it has developed a corky layer, plant each piece in a pot or in the open ground. Then label.

PROPAGATING FROM CORMELS

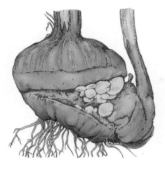

Cormels or cormlets are miniature corms that sometimes form between the new corm and the old, disintegrating corm. The number produced can vary widely, and will increase the deeper the corm is planted – a gladiolus corm may produce as many as 50 cormels.

To use cormels for propagation, lift the parent corm at the end of the growing season and separate all the cormels. Store them in an environment that is dry, well ventilated and frost free but cool – below 5°C (41°F). The following spring plant them out, soaking first for 24 hours if they have dried out. Cormels are likely to take at least two years to reach flowering size.

in the soil with a trowel or hand fork and gently lift, separate and replant at the original depth. Chains of corms can be replanted intact or individually separated; there is evidence that maintaining the chains intact may be the better option.

On small corms, or those that have few buds, you can remove the main stem by snapping it off or cutting it out to induce side buds to form shoots. Dust all the cut surfaces with a fungicidal powder. Dig a hole twice the depth of the corm in the open ground and plant it out immediately and label. At the end of the growing season, you will have several small corms or a large corm with several buds.

Another way to propagate corms artificially is by cutting them into several smaller sections while dormant. Having dusted each cut surface with fungicidal powder, label the batch clearly. Then leave the pieces for 24–48 hours in a warm, dry place (18–21°C/65–70°F) to allow the cut surfaces to seal. Once the pieces have developed a corky layer over the cut surfaces, plant each singly in small pots of multipurpose potting compost. Grow them on until large enough to plant out in the garden or pot on if the corms are to be grown indoors.

• TUBERS •

True tubers, as opposed to tuberous roots (see p82), are enlarged structures that develop on modified underground stems. They store food and nutrients to help the plant survive dormant periods in winter or drought conditions and provide energy for regrowth during the next growing season. Tuber development is encouraged by reduced daylength and lower night temperatures. Tubers contain dormant buds, usually referred to as 'eyes'.

Ornamental tuberous plants include *Caladium*, winter aconite (*Eranthis*) and flame creeper (*Tropaeolum speciosum*); and some waterlilies (*Nymphaea*) produce tuber-like structures.

Although potatoes (*Solanum tuberosum*) are prolific producers of tubers, this is unusual. Most tuberous plants produce only a small number of tubers, but you can propagate more plants by division.

Before growth starts in spring or just as the eyes are breaking into growth, divide the tuber into pieces with preferably two or three buds, or eyes. Leave the pieces in a warm, dry place for 24–48 hours. Once the pieces have developed a corky layer over the cut surfaces, plant them out singly in small pots of multipurpose compost, growing them on until large enough, or else plant out directly where you want them to grow. Do not keep them in the dry for any longer than is necessary.

POTATO TUBERS

Tubers are storage organs that develop as roundish, swollen areas on modified underground stems, usually at the tips.

TUBERCLES

A few plants produce miniature tubers (called tubercles) above ground in the leaf axils. Once mature, they fall off and develop into new plants. Better still, carefully remove them – they come away easily when mature – and plant straightaway in a place of your choice, at twice their own depth.

Achimenes (hot-water plants) produce tubercles.

DIVIDING TUBERS

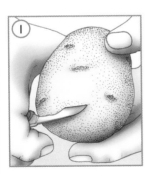

Cut a tuber into pieces with a sharp knife. Ensure each piece has at least one good bud, or 'eye'.

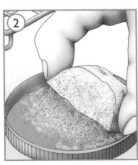

Dust all cut surfaces with sulphur. Stand on a wire rack in a warm (21°C/70°F), dry place.

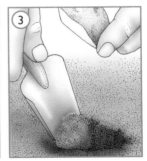

Once the tuber pieces have formed a protective corky layer, plant them. Label them clearly.

• RHIZOMES •

A rhizome is a perennial modified stem that grows more-or-less horizontally on or just below soil level. Because it is a modified stem, it produces new leaves and shoots and has its own roots, or readily develops them when separated from the parent plant. This makes rhizomatous plants easy to propagate.

The rhizomes of some plants – those of bearded irises, for example – are large and are used to store food reserves and nutrients. Others, such as bamboo and mint (*Mentha*), are much thinner and longer and, although they usually store some food, they are mainly used by the plant to spread and cover new ground.

Rhizomes spread outwards, and then the oldest parts die off in the centre. To keep plants healthy and maintain vigour and flowering, regularly lift and divide them, discarding the oldest rhizomes and replanting only the youngest sections.

You can propagate a number of plants by dividing their rhizomes. Most evergreen rhizomes are lifted and divided after flowering. Deciduous rhizomatous plants are divided in spring just before new growth begins.

Having dug up the rhizome, remove the excess soil – this is easier to do when the soil is dry. Shorten the leaf blades to help reduce windrock and water loss.

DIVIDING RHIZOMES

Divide iris rhizomes after flowering, when the old root system dies down and new rhizomes start to form.

Carefully lift a clump with a garden fork. Knock off as much soil as possible from the roots.

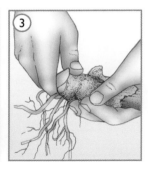

Cut away and discard any old or damaged rhizomes, leaving the current season's new growth.

Replanting irises too deeply will prevent the plant from flowering for a couple of years until the rhizome raises itself to soil level.

CROWN RHIZOMES

Less easy to propagate are the 'crown' rhizomes such as peony and asparagus, which have what is traditionally referred to as herbaceous perennial rootstock.

Lift crown rhizomes when dormant, such as in late winter, before the buds enlarge and before the roots start to develop. Using a spade, divide the rootstock into suitable pieces, ensuring each one has at least one developed growth bud – preferably two or more.

Treat the cut surfaces with sulphur powder and replant once this has dried – sooner rather later, especially with peonies – and at the same depth as originally growing.

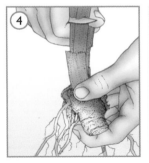

Prepare the rhizome by shortening the leaf blades or stems and cutting back the roots to 5–7.5cm (2–3in).

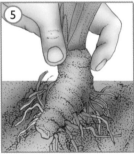

Replant each rhizome at the original depth on a ridge with its roots spread out in trenches either side.

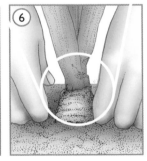

Firm back the soil over the roots. Barely cover the rhizomes of irises. Settle in by watering the soil.

DIVISION

• INTRODUCTION TO DIVISION •

Dividing a plant is a common way to propagate many herbaceous perennials and it is also used to rejuvenate plants. Propagation by division is also successful with a few shrubs that produce dense clumps of stems. Cut back the stems to reduce water loss, then lift the plant when dormant and divide the clump into suitable portions, but discard any old sections with few roots. Replant in well-prepared soil at the same depth as the plant was originally growing. You can also divide shrubs that produce suckers (see p75).

For most perennials, the best time for propagation is immediately after flowering or for late-flowering plants in autumn or the following spring, but there are exceptions. Plants with fibrous crowns can be divided when new shoots are just beginning to form. Many herbaceous perennials, such as astilbe, hosta, red hot poker (*Kniphofia*) and meadow rue (*Thalictrum*), produce compact, fleshy crowns that are not easy to pull apart. The best time to divide these is towards the end of their dormant period, when their buds begin to shoot and you can more easily select the most suitable sections of the crown.

The most common way to divide herbaceous perennials is used for

DIVIDING A HERBACEOUS PLANT WITH A FLESHY CROWN

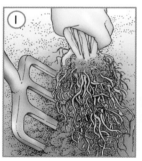

Towards the end of its dormant season, as new shoots appear, carefully lift the plant to be divided.

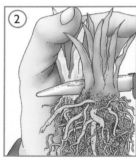

Cut the crown into suitably sized pieces, ensuring each one has at least one well-developed bud.

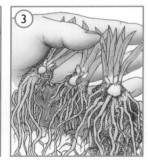

If necessary, dust the cut surfaces with sulphur powder. Then replant immediately.

those plants that have fibrous roots and a relatively loose crown. As the crown develops it produces the newest, strongest shoots at its edges and these are separated and planted up.

As well as increasing perennials with fibrous or fleshy crowns, division can also be used for propagating plants with tuberous roots (see p82), tubers (see p92) and rhizomes (see p94).

DIVIDING A HERBACEOUS PLANT WITH A FIBROUS CROWN

Carefully lift the plant that is to be divided, preferably when dormant or new shoots are being produced.

Shake off as much soil as possible. Wash the crown and its roots and remove any badly damaged roots.

Shorten all tall stems and excessive growth to minimise water loss while the divisions are establishing.

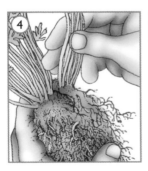

Carefully break or cut off sections from the edge of the crown, ensuring each one has good buds or shoots.

Replant the divisions at the original depth in well-prepared soil straight away. Firm the soil and label.

Water in thoroughly to settle the soil around the roots, using a watering can with a rose.

• OFFSETS & RUNNERS •

Some herbaceous perennials produce either offsets or runners (types of modified stems) that can also be propagated by division. Offsets and runners are similar as they are both types of plantlets that develop naturally on the ends of stems arising from the parent plant. The major difference is that offsets tend to have shorter, thicker stems. They are also often slower than runners to produce their roots.

OFFSETS

These plantlets develop on the ends of sideshoots growing above or below ground on such plants as some saxifrages, houseleek (*Sempervivum*) and phormium.

Offsets generally are miniature versions of the parent but are very dependent on the parent plant for food and energy because initially they have few roots of their own. Roots usually develop towards the end of the growing season and have to be encouraged by severing the connection with the parent. When removing a young offset cut the stem close to the parent plant to prevent rot setting in. Once the offset has developed a good root system, you can plant it out.

The term offset is also used to describe bulblets and bulbils (see p86) and shoots that grow from the roots of many plants, mainly monocotyledonous ones such as cabbage palm (*Cordyline*) and yucca.

PROPAGATING BY OFFSETS

Carefully remove a healthy young offset, either by pulling it away from the parent plant or by cutting with a sharp knife.

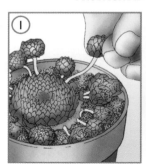

Plant each offset in pots filled with cuttings compost with added grit. Ensure their base is just in contact with the compost.

DIVISION

PROPAGATING BY RUNNERS

In spring or early summer thin out some runners to encourage strong new growth.

Dig a hole beneath a plantlet and set a pot of compost in it. Pin down the plantlet using a wire staple.

Pin down as many plantlets as needed in a star-shaped pattern around the parent plant.

These can be removed by scraping away soil at the base of the plant and cutting away the offset with a knife, preferably with a section of root attached.

If a plant produces few offsets, you can try severing its growing tip to encourage offsets to form. Most offsets can be removed in their first year, but some such as yuccas grow slowly and should be separated only when they are a few years old.

RUNNERS

Runners are generally the horizontal stems that are produced from the main plant and then creep along the ground. They root and produce plantlets along their length and/or at their ends. This is how *Geum reptans*, herbaceous potentillas, strawberries and some grasses are propagated.

When propagating from runners, it is better to have a few larger plants than lots of small ones, so select only runners with vigorous growth.

Secure the runner in a suitably sized pot filled with good-quality potting compost. If rooting plants start to produce their own runners, carefully cut these off to prevent the plant putting its strength into the runner and developing further plantlets.

Sever the connecting stem when the plants are fully rooted, and plant out.

LAYERING

INTRODUCTION TO LAYERING

Layering is one of the oldest methods used to propagate woody plants. A stem is encouraged to produce roots while still attached to the parent plant, which feeds it during the rooting process. The layer is then removed once it has rooted. There is usually no need for complicated environmental control, and outdoor plants can be layered *in situ* in the garden.

A new plant is nearly always guaranteed when increasing a plant by layering, which is one of the easiest but slowest propagation methods, usually taking 12 months or more.

As with all methods of stem propagation, the youngest, most vigorous shoots have the best ability to root. These are the ones that should

PROPAGATING BY SIMPLE LAYERING

During late winter or early spring cultivate the soil around the plant. If it is needed, add compost and grit to improve the soil and help good rooting.

Some 15–30cm (6–12in) from the growing tip of a young, vigorous stem trim off the leaves and sideshoots for 10–30cm (4–12in).

Bring the leafless part of the stem down to ground level and mark its position. Dig a shallow trench at that point. If necessary, girdle the stem (see p109).

Carefully bend the stem at right angles where the stem meets the trench. Then peg it down in the trench so that the leaves at the tip are just above the ground.

be used or be encouraged to form by cutting back the plant hard well before propagation is to take place.

As roots form in the dark, it's important that the area where roots develop is kept as dark as possible. Warmth will also help the rooting process, so start taking layers from midspring and if layering outdoors try to layer in a sunny position.

SIMPLE LAYERING

This layering technique is an effective propagation method for shrubs that do not root readily from cuttings. It is most suitable for plants that have stems that can be bent down to ground level. If this is not possible, try increasing your woody plants by air layering (see p107).

Simple layering can be carried out in autumn or spring. Deciduous plants

Return the soil to the trench so the stem is buried completely. Firm it in well. Water well and keep the soil moist, especially in dry periods.

Once the layered stem has rooted, which can take 12 months, sever it from the parent plant in autumn. Leave it to establish as a new plant.

About 3–4 weeks later cut off the growing tip from the rooted layer. This will help to divert the plant's resources into the roots.

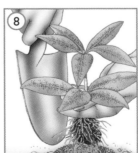

Once its roots are well established, lift the layered stem, or leave for a further year. Replant in well-prepared soil or in a pot.

You can propagate several new plants from a short length of climber stem, such as ivy.

respond well in either season, but evergreens respond better in spring. Choose flexible young shoots on the outside of the plant. Applying hormone rooting powder to the cut on a girdled stem (see p109) may aid rooting.

When rooting into garden soil, improve it by digging in plenty of organic matter, to hold moisture, as well as sharp sand or grit if it is heavy clay soil and poorly drained. Make a shallow trench, 10–15cm (4–6in) deep, in the soil and peg the wounded section of stem into the bottom, using metal pins. Alternatively, sink a 10–15cm (4–6in) pot filled with good-quality John Innes seed compost into the ground and peg the stem into that. Weigh down the stem at soil level with a stone or large pebble; this will also help create a dark rooting environment.

Secure the bit of stem that is above ground by tying it to a vertical cane so that it grows upwards.

• AIR LAYERING •

Air layering is useful for difficult-to-root shrubs whose stems cannot be bent down to ground level for simple layering (see p105). It can also be used for indoor plants like fig (*Ficus*) and dracaena and where there is no space for a simple layer to root in the ground. Air layering is usually carried out in spring, on the

PROPAGATING BY AIR LAYERING

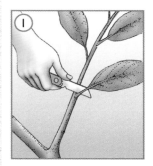

Trim leaves and sideshoots from a 15–23cm (6–9in) length some 30cm (12in) from the tip. Girdle the stem.

Treat the girdle with hormone rooting powder, then cover the area with a large ball of wet moss.

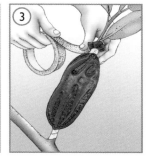

Hold the moss in place with a square of black polythene. Secure with sticky insulating tape.

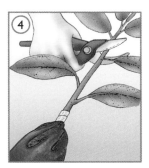

Once it has rooted, prune back any new growth on the layer towards the end of the dormant season.

Cut the stem just below the point of layering, using a pair of secateurs. Remove the black polythene square.

Loosen the moss ball and roots slightly. Then pot up and place out of direct sunlight until established.

LAYERING

OTHER LAYERING METHODS

There are a couple of other modified layering methods you can use when you need a lot of plants.

Serpentine layering involves looping the stems of climbers in and out of the soil to encourage roots to form at several points along the stem. The technique for each buried section is the same as for simple layering (see p105). Thin-stemmed climbers do not need to be wounded. This method is suitable for climbers with long, flexible stems, such as clematis, *Hydrangea anomala* subsp. *petiolaris*, hop (*Humulus lupulus*), honeysuckle (*Lonicera*), Virginia creeper (*Parthenocissus*), *Vitis* and wisteria.

French layering involves cutting back the parent plant hard in spring to produce lots of new stems near ground level. The following spring, peg these new shoots down to the soil like the spokes of a wheel radiating out from the base of the plant. As sideshoots grow upwards from these stems, gradually mound soil over them to a height of 5cm (2in), to encourage rooting. By the autumn or the following spring, these rooted sections can be separated and planted out. French layering is suitable only for shrubs that grow vigorously, such as dogwood (*Cornus*), smoke bush (*Cotinus*) and *Prunus tenella*.

previous year's growth, or in the late summer on the current year's growth. It usually takes at least a full growing season to produce a good root system.

Girdle the stem at a leaf node (see box right). The best methods of girdling when air layering are cutting the stem upwards at an angle from below or removing a complete ring of bark.

Soak some sphagnum moss so that it is thoroughly wet, preferably overnight. Knead two large handfuls into a ball 8cm (3in) across and then divide into two with your thumbs. Place the two pieces around the treated area of the stem and knead the moss together again so that the ball stays firm. Secure black polythene around the moss so it excludes light at both ends, retains moisture and maintains warmth.

Once roots have grown through the moss, remove the rooted stem and pot it up into a pot filled with multipurpose or John Innes No 1 compost. Gently firm in. Keep the new plant in a sheltered place until it becomes established.

STEM GIRDLING

Old stems or the stems of old plants may not root easily. If that is the case, it may help to girdle the stem – where the stem is deliberately wounded. This either impedes or stops the flow of sap and so nutrients to the rest of the branch, and this in turn encourages rooting.

Rubus phoenicolasius – Japanese wineberry – can be propagated by layering.

WAYS TO GIRDLE A STEM

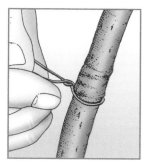 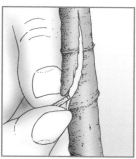 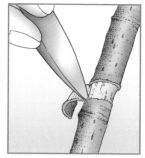

Secure a piece of copper wire round the stem where it is to root. Twist it so that it is finger tight.

Make an angled cut half way through the stem. Keep the cut surfaces apart with a matchstick.

Remove a ring of bark, about 6mm (¼in) wide, from round the stem, using a sharp knife.

• TIP LAYERING •

This technique works well for blackberries, hybrid berries and other *Rubus* species. It is a simple process – one that wild species do naturally and is the method by which blackberries colonise new ground. The tip of a long, arching stem that easily reaches ground level is buried 8cm (3in) under the surface of the soil, in mid- to late spring. If necessary the stem is held in position with a loop of thick wire. Roots should have developed from the shoot tip by autumn or the following spring.

However, better results will be achieved if you take a few precautionary steps, such as preparing the propagation material, before increasing your plant by tip layering. Select a new, strong stem in spring and, once it reaches up to 45cm (18in) long, remove the tip. Continue to remove the tips; by midsummer six to eight sideshoots should have developed.

Pull down the stem to ground level and mark where it touches the soil. Start digging at this spot, to a depth of 15cm (6in), working in bulky organic matter and, if necessary, grit to improve structure and drainage. Then dig a sloping trench, 10cm (4in) deep at its deepest end. Give it smoothed vertical sides except for one which should be sloped towards the parent plant.

PROPAGATING BY TIP LAYERING

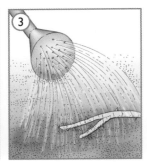

Pinch out the tips of basal shoots to encourage new growth. Cultivate the soil close to the plant.

Pull down a stem and dig a sloping trench where its tip touches the ground. Pin it down with a staple.

Completely cover the tip of the shoot with soil, firming it well and watering thoroughly.

Smoothed vertical sides will help any shoot to grow vertically and so produce a manageable plant. Place the stem tip in the deepest part of the trench and pin down with a strong wire staple or hold the stem in place with a stone.

In autumn cut the main stem close to the crown of the parent plant to encourage each layer to establish as an independent plant.

Once the new layer has dropped its leaves, cut off the remainder of the original layered stem, close to the newly layered plant at ground level. Shorten some of the topgrowth from the new plant. Then lift it and replant in its final position immediately or pot up with a good-quality potting compost in a suitable-sized pot.

Propagation by tip layering in this way can be repeated each year.

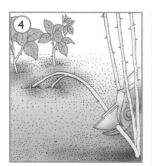

In autumn cut back the original stem close to the parent plant to encourage further rooting.

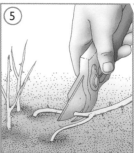

Once the rooted layer has dropped its leaves cut off the rest of the stem. Shorten some of the new topgrowth.

Lift the layer very carefully to avoid damaging its roots. Plant it at once in well-cultivated soil.

• DROPPING •

Dropping is an unusual but easy method of propagating a number of woody or shrubby plants. This includes heathers, lavender and rosemary, dwarf rhododendrons, kalmia and gaultheria. It is a good way of rejuvenating straggly, overgrown or, sometimes, worn-out plants – although these may not be

PROPAGATING BY DROPPING

In the dormant season prune a plant to induce lots of strong, new shoots, which will root easily.

The following spring cultivate the soil, adding compost and grit if necessary. Dig a very large hole.

Lift the plant with as complete a rootball as possible and carefully place it in the hole.

Leave only 2.5cm (1in) of the stem tips exposed when covering the plant with soil. Firm in.

During the growing season keep the soil moist by watering if it starts to dry out.

In autumn lift the plant, cut away each rooted stem and plant it out or pot it up individually.

LAYERING

vigorous enough to root well. Often the plants resulting from dropping are less shapely and so less desirable than those produced from cuttings.

If the plant will tolerate propagation by dropping, prepare it well in advance by cutting it back as hard as possible to encourage strong, new stems that will have more potential to root. Then, in spring, dig a hole wide enough and deep enough so that when the whole plant is 'dropped' into it only the top 2.5–5cm (1–2in) of stems is visible.

Backfill the hole with well-prepared soil and arrange the stems as you do so in the most suitable positions, depending on the habit of the plant (see box above). Make sure you work soil around the stems to prevent any air gaps.

During the summer give liquid feeds to encourage good growth.

In autumn, carefully remove the soil from around the stems or lift the whole plant. The branches should have rooted, usually close to the soil surface. Cut away each rooted stem and either plant out or, preferably, pot up individually and grow on until big enough to plant out the following spring. The old plant should then be discarded.

WAYS TO BURY A PLANT

The traditional method is to push out all the branches to the perimeter of the hole and fill in the middle of the hole with soil. If the branching is sparse, compress and arrange the branches into a single row near the middle of the plant. This pattern is not suitable for plants that produce a thick mass of stems as there is insufficient room for root development. For plants with brittle branches, simply work the soil down among them as they lie.

MOUNDING UP

If lifting the plant and dropping it into a large hole are not suitable, instead try mounding up a mixture of soil and compost around the plant's stems. Rooting will not be as good if the soil is allowed to dry out, so it may be better to make a bottomless box out of wood or similar material, position it around the plant and then fill this with the soil/compost mix.

SEEDS

• INTRODUCTION TO SEEDS •

Seed sowing is one of the most satisfying and cheapest methods of propagation, but it can be frustrating when seeds fail to germinate or the resulting seedlings struggle to develop properly. This may be due to a number of reasons (see p132).

The seeds of hardy trees, shrubs, perennials and annuals can be sown outside where you want them to grow. Others such as half-hardy annuals and half-hardy bedding perennials will need to be sown indoors, preferably with bottom heat if you want good results (see p126).

Some seeds readily self-sow, and you can carefully lift their seedlings, potting them up and growing them on until they are larger, well-established plants, or just replanting where you want them to grow in the garden.

SEED MORPHOLOGY

Seeds contain an embryo (miniature plant), a dried food supply and a protective outer seed coat. The seed may be enclosed within a fleshy fruit or a tough outer skin.

A seed's embryo consists of an undeveloped shoot attached to seed leaves (cotyledons), an undeveloped root and a short section of stem between them. When the seed germinates the seed leaves remain in the soil or emerge above ground with the seedling. They are usually simple in shape and quite unlike the characteristic mature (true) leaves. The seed's food reserves may be stored in the seed leaves or in a separate food store.

Most seeds store their food reserves as dried carbohydrates; they generally survive longer than those seeds that store their reserves as fats or oils, such as sweet chestnut (*Castanea*), magnolia and peony. Such seeds do not like to be dried and stored and so should be sown immediately after harvesting.

OUTDOOR SEED SOWING

Many hardy annuals, biennials and herbaceous plants can be sown directly outdoors where you want them to grow. Most seeds need a minimum temperature of 7°C (45°F) for germination, but there are many exceptions. Always refer to the seed packet for the best time to sow, as it does vary with plant type.

It is hugely satisfying to produce new plants from seed and to nurture seedlings into mature plants.

SEEDS

It is also important to prepare a good seedbed, free of weeds and with a fine, crumbly soil tilth. Do this in advance to allow time for the soil to settle. Dig in plenty of well-rotted organic matter if the soil conditions are poor, and even sharp grit in heavy clay soils. Use a rake to level the surface and create a crumbly tilth. Then pick off any weeds and debris.

Hardy annuals can be sown directly outside where you want them to grow and flower.

GENETICS

Plants grown from the seed of species will be similar to the parent plants – that is, come 'true to type'. If, however, one of the parents is genetically unstable or there is cross-pollination with other species or types, the new plants will be more variable, although they will have the majority of the desirable characteristics of one or both parents. Such seeds are still worth sowing as the variability is usually quite small.

Plants grown from seed of F1 cultivars will be extremely variable. The process of producing the F1 seeds in the first place is very complicated and involves creating pure breeding lines with specific characteristics required for the resulting seedlings. Trying to reproduce from such 'artificial' seed is generally not worthwhile.

Seed can be sown by one of two methods: broadcasting or drilling. Broadcasting involves scattering the seed thinly over the soil surface; it is often a better method when seeding large areas, but when thinning out the seedlings it can be difficult to differentiate between the cultivated plants and weed seedlings. Aim for a gap of about 6mm (¼in) between seeds. Cover the seeds lightly with soil by carefully raking over the area. Water well with a watering can fitted with a fine rose.

Drilling (sowing in drills or rows) is a more precise method, and thinning out is easier: you know where the seeds have been sown so can remove all other seedlings safe in the knowledge they are weeds. Use a bamboo cane, small hoe or the corner of a rake to produce a straight drill (shallow depression) of a depth appropriate for the species being sown. Space the drills according to the eventual plant height; instructions are usually given on the seed packet.

Before sowing water the bottom of the drill, then thinly scatter the seed in the drill – 6mm (¼in) apart is usually adequate. Use a rake gently to cover the seeds with soil, filling the drill back in again. Place a label in the soil at one end of the drill. If necessary, cover the area with a single layer of fleece. Use a spade to push the edges of the fleece into the ground to ensure it doesn't blow away. Remember to water in dry spells.

• COLLECTING & STORING SEEDS •

One of the most satisfying ways of growing plants from seed is by using seed saved from plants in your own garden or in a friend's or neighbour's garden.

Although you can save seed from just about any plant, those from highly bred F1 hybrids are not worth saving (see box on p119). Similarly, saving and sowing pips and stones from tree fruit cultivars will usually result in plants that produce little and/or poor-quality fruit.

It is important to label seeds at all stages of the collecting, drying and storing processes, as there is nothing worse than having a packet of seeds with no idea as to what is in it.

Collect seed once it is completely ripe, but before it is dispersed from the plant

DRYING FLOWER SEED FOR STORING

Spread fleshy capsules on kitchen paper in a tray or box. Leave to dry in the sun or a warm place.

Bunch flower stems together before hanging them to dry with their heads enclosed in a paper bag.

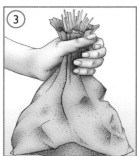

Tie the neck of the bag and hang it in a dry, airy place. Shake occasionally so the seeds drop into it.

Label each paper packet with the plant name and date. Then place cleaned, dried seeds inside. Store in a cool, dry, dark place, such as a fridge, in an airtight container.

– sometimes a very fine line, so keep an eye on plants as the seedheads mature. Whenever possible, do this on a fine, dry day. Before storing, dry the seed and remove all debris by picking it out, sieving or gently blowing it away.

Large, fleshy seeds such as cyclamen, lilies and hellebores do not respond well to drying and should be left to mature on the plant, to be collected just before they are dispersed; place a paper bag over the seedheads to prevent dispersal.

Seed enclosed in berries or fleshy fruit can also be collected, but you have to take care to remove all the fruit and properly dry the seed. Again, leave the fruit on the plant until it is fully mature, or the seed will not be ripe enough to store and germinate properly. Having removed as much flesh as possible, place the remaining seed mass in a jar of tepid water and leave in a warm place for up to 24 hours. Then pick off any remaining skin by hand. Place the seeds on kitchen paper to dry thoroughly.

STORING

Although the vast majority of seeds can be stored, there are some seeds such as anemone and snowdrops (*Galanthus*) that quickly lose their viability so do not store well and have to be sown fresh.

VIABILITY TEST

To check whether seed is still viable, take 20 seeds and place on damp kitchen paper or blotting paper on a saucer. Cover with plastic wrap and place somewhere warm, such as in an airing cupboard. Check daily and count the number of seeds that have germinated. Multiply this number by five to give the percentage that will germinate – that is, the germination rate. Most bought seed has a germination rate of 80 percent or more. If your seed has a lower rate than this, it is worth sowing more thickly; if it has a very low rate it may not be feasible to sow at all.

One of the easiest ways to store large seeds or large amounts of one type of seed is to use a small, airtight, plastic or glass bottle or jar. Alternatively, place small amounts of seed in paper packets and seal in an airtight plastic box. To ensure the seed remains dry, store them with sachets of silica gel, which will absorb moisture from the atmosphere.

Most seed stored properly can remain viable for at least two or three years.

• BREAKING SEED DORMANCY •

The seeds of many trees, shrubs and other woody plants and those of some herbaceous perennials have a natural dormancy to help them survive hostile environments and germinate only in favourable conditions, such as the warmth of spring, or after rain. Gardeners can supply or manipulate favourable germination conditions just about all year round, and in order to do this they sometimes need to break this dormancy artificially for seed to germinate.

Seed dormancy can prevent the seed from germinating for two or more years. It is therefore necessary to intervene so that seeds can be sown and allowed to germinate without having to wait this long.

There are a number of different techniques for breaking dormancy, depending on the plants and the conditions you can provide.

To enable seeds with hard seed coats to absorb the water needed for germination you can scrape them, cut them or soak them before sowing. Examples of seed benefiting from these scarification treatments include plants from the pea family, such as bladder senna (*Colutea*), broom (*Cytisus* and *Genista*), bush clover (*Lespedeza*), lupins, robinia and Spanish broom (*Spartium*).

If tree and shrub seeds fail to germinate in their first season after sowing, it is worth leaving them for

BREAKING DORMANCY BY SCRAPING

Take a jar with a screw-top lid and line it with a sheet of sandpaper. Place some seeds in the jar and screw on the lid.

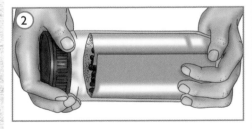

Gently shake the jar until the seed coats are worn down, thereby enabling the seeds to take up water.

BREAKING DORMANCY BY CUTTING OR CHIPPING

You can cut or chip seeds that have a hard seed coat with a sharp knife or razor blade. Be careful, when doing this, to prevent cutting yourself. Always cut away from your body. The aim is to nick or remove just a small section of the seed coat, preferably near the embryo scar, which is where water is absorbed. Always cut shallowly, to avoid cutting into the embryo.

another season to see if they perform, because germination can often be erratic and slow, especially for seeds where dormancy is a problem.

BREAKING DORMANCY BY HOT WATER TREATMENT

Prepare seeds with hard or impermeable seed coats for sowing by treating them with hot water. This allows the seeds to take up water and swell so they can germinate. Place the seeds in a shallow dish.

Using a ratio of three parts water to one part seeds, pour water that has just gone off the boil over the seeds. Place the dish in a warm environment and leave for 24 hours. If the seeds do not swell, repeat the exercise.

SCARIFICATION

A hard seed coat can be rubbed with a file until it is sufficiently worn away or be chipped until the seed is exposed and water can be taken up (see box on p123). Better still, especially with small seeds or where you have a lot of seeds to treat, use a sheet of sandpaper; the grade used depends on the size of seed and the thickness of the seed coat (see p122).

BREAKING DORMANCY BY COLD STRATIFICATION

Sieve four parts by volume of leafmould, composted bark and/or coir (or a mixture) to one part seeds.

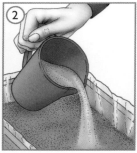

Add sufficient water to the material so it exudes a little water when squeezed lightly in the hand.

Measure out one part by volume of seeds and sprinkle evenly over the surface of the moist material.

Mix all the ingredients thoroughly. Add one part by volume of sharp grit if the mixture looks compact.

Place the mixture in a polythene bag. Label and leave in a warm area for two to three days.

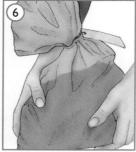

Move the polythene bag to the fridge to chill; this can take up to 20 weeks. Turn and shake occasionally.

Hot-water treatment

Some seeds benefit from soaking in water. Although cold water will help, warm or even hot water is usually more effective. This removes 'waterproofing' material from the seed and allows it to take up water. Care should be taken when soaking seed, as soaking for too long can lead to rotting.

Usually, the seeds are ready to sow when they sink; seeds that continue to float are non-viable and should always be discarded.

STRATIFICATION

This is used for seeds that respond to either heat or cold and overcomes the biochemical control of the dormancy. Stratifying will break dormancy by mimicking the conditions these seeds are naturally subject to in the soil that trigger germination. Most commonly this is a spell of cold temperatures similar to that experienced in winter.

Plants used to cold winters may need a period of chilling at 1–3°C (34–37°F) to break their dormancy. Often, simply sowing in autumn and placing the pot of seeds outside is enough to stimulate germination. You can also store and chill seeds in the fridge over winter for a spring sowing.

Cold stratification

Sift leafmould, composted bark or coir through a coarse sieve. By volume, you will need four times the amount of this for one part of seed. Moisten with tepid water. Mix in the seeds, introducing plenty of air as you do so. If the mixture looks too dense add grit or vermiculite. Transfer the mixture to a plastic bag.

Seal and label the bag and leave in a warm place for two to three days. Chill the seeds in a fridge, kept below 5°C (41°F) for four to twenty weeks. Check regularly and pot up any germinated seeds. Most seeds will not germinate until sown in warmer conditions.

Warm stratification

A few seeds such as ash (*Fraxinus*) have multiple dormancy and germinate only in their second spring, having been subjected to a warm spell followed by a cold period. This is done by placing the seeds in a plastic bag (as for cold stratification) but they are kept at 18–24°C (65–75°F) for up to 12 weeks, before being moved to a fridge, for chilling.

Alternatively, sow in pots and place in a heated propagator for the required period. Following this, move the pots into a cold frame for the winter.

• SEED SOWING IN CONTAINERS •

Sowing seeds indoors in containers allows you to start off tender plants, such as half-hardy annuals, earlier in the season and gives them the protection they need from cold outdoor conditions. You can also sow seeds of hardy plants in containers, as this will provide young plants sooner than sowing outside and so extend the growing season.

If you have a heated glasshouse, a heated propagator or enough space and light in your home, such as a conservatory or well-lit windowsill, you can sow plants that need a long growing season: for example, snapdragon (*Antirrhinum*), *Begonia* and geranium (*Pelargonium*) can be started into growth as early as midwinter.

The seeds of most plants are sown indoors in late winter or spring, for planting out in late spring or early summer when there is little risk of frost. But it is possible to sow any plant at its correct sowing time indoors, providing you can establish the necessary growing conditions. Indoor sowing also prevents any possible damage from the vagaries of weather or pests.

Before choosing a pot, pan or celltray or seedtray decide how much seed is to be sown. The container should be large enough to allow the seedlings space to develop to the size at which they are able to be pricked out. Small seeds are generally sown in small pots or seedtrays and pricked out (transplanted) into larger pots at the seedling stage. They and larger seeds can be sown into celltrays or soil blocks (see p24), which limit transplant shock for those plants that dislike root disturbance. Large seeds can also be sown individually into 8–9cm (3–3½in) pots; it is usually a good idea to sow two seeds per pot and then remove the weakest seedling if both seeds germinate.

Containers should be clean and compost should be fresh and provide adequate aeration, sufficient water-holding capacity, a neutral acidity/alkalinity reaction and sufficient phosphate. Thus a 'sowing' or 'seed' compost should be used (see p26).

For best results, the compost surface should be smooth, flat and level. Therefore sieve very coarse compost to remove large lumps, which will affect the overall structure of the compost and reduce germination.

It may be necessary to soak large seeds in water for up to 24 hours before being sown, so they take up moisture more easily, which will assist with germination and the establishment of a new plant.

Having filled the container with compost, tap the container at the side or knock it gently on the work surface to help settle the compost. Gently firm with your fingertips to eliminate air pockets, but do not overfirm. Add more compost if necessary, and tap the container again to settle it.

SOWING SEEDS IN CONTAINERS

If necessary, soak large seeds in a bowl of tepid water for 12–24 hours before sowing.

Fill a container with compost until it is heaped above the rim. Then tap the container to settle the compost.

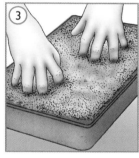

Gently firm the compost into the corners and base using the tips of the fingers.

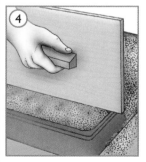

Level off the excess compost with a presser board, using a sawing action until it is level with the rim.

Use the presser board to firm the compost lightly to 0.5–1cm (¼–½in) below the rim.

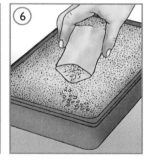

Sow half the seeds evenly across the container, keeping your hand low to prevent the seeds bouncing.

Level off the excess compost to the rim of the container with your hand, a compost tamper or a container of the same size.

After sowing, cover the seeds to no more than half their own depth with sieved compost, then water them with tepid water (see p30). Protect the container with a piece of glass or plastic wrap, which can be left in place until the seeds germinate, or set in a propagator. Place a sheet of paper over the glass or

SOWING SEEDS IN CONTAINERS (CONTINUED)

Turn the container through 90 degrees. Sow the remaining seeds evenly across the surface.

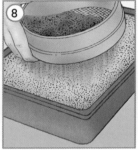

Lightly cover the seeds by sieving over compost, keeping the sieve low over the seeds.

Label the container with the plant's full name and date of sowing. Soft pencils are often best for this.

Water in the seeds from above using a can with a fine rose. Or stand in a tray of tepid water.

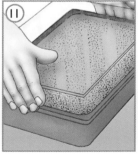

Cover the container with a pane of glass to keep the seeds moist and warm.

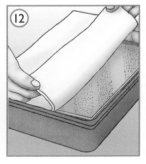

Place a sheet of paper over the glass to minimise temperature fluctuations.

plastic wrap to keep out strong sunlight and reduce temperature fluctuations, or cover the propagator with fleece.

Stand the container in a warm, dark place such as an airing cupboard, or set the thermostat on the propagator, depending on the germination temperature requirements of the particular plant; a temperature of 18–21°C (65–70°F) suits most plants. Even temperatures are important for good germination and growing on. If the temperature fluctuates widely, germination will be erratic and seedling growth affected. The easiest way to ensure an even temperature is to use a thermostatically-controlled propagator.

As long as viable seeds are used and the correct conditions are given, germination can take place in as little as 5–7 days. The higher the compost temperature, the quicker germination occurs, providing it does not reach critical temperatures. Forcing germination in very high temperatures can lead to weak, leggy seedlings, especially if light levels are too low. As there is a correlation between temperature and light levels for strong growth, in these situations it is better to reduce the germination or growing-on temperature around the containers.

SOWING TECHNIQUES

Some seeds need specific sowing treatments, such as light exclusion or their dormancy broken by stratification or scarification (scraping or chipping the seed coat, see p122), so check for particular instructions.

Small seeds are broadcast, or scattered thinly, over the surface of the compost. Very small seeds can be mixed with fine, dry sand before sowing to obtain an even distribution. After scattering, sift a layer of compost or fine vermiculite over the seeds.

Larger seeds can be station sown: each seed is pressed individually into the surface of the compost 1–2.5cm (½–1in) apart. If using a celltray or soil block, sow one seed per cell or insert two seeds and after germination discard the weaker seedling.

Sowing alpine seeds requires special consideration. Generally, these are sown in autumn or winter into containers that are then stood unprotected from the elements. As the seedlings may not be big enough to handle until the following year, the kind of compost in which they are sown is important as it will have to sustain the seedlings for at least 12 months. It should also contain a high proportion of grit for good drainage.

• CARING FOR SEEDLINGS •

Although the germination stage is usually the trickiest part of producing young plants from seed, the seedlings also need careful nurturing if they are to produce strong, healthy plants that perform well.

Once germination occurs, remove the glass, polythene or propagator lid on warm days to give seedlings plenty of air, but replace it whenever temperatures are likely to drop; most seedlings produced in warm conditions need a minimum growing-on temperature of 10–13°C (50–55°F) until well established.

Seedlings need good, even light for strong growth. Never expose them to harsh, burning sunlight; if necessary, shade them with horticultural fleece. Even, all-round light is important, otherwise the seedlings become long and drawn and will never perform well. This is usually only a problem when growing on a windowsill.

Transplant the seedlings once they are large enough to handle, to give them more room to develop in. When pricking out (transplanting) always hold seedlings by their seed (first) leaves between thumb and forefinger, rather than by the stem; if you damage the stem, the plant will never develop properly and will more than likely die.

CARING FOR SEEDLINGS

Spray seedlings regularly with water, but do not allow the compost to become waterlogged.

To transplant, use a dibber and gently lift a seedling by holding its seed leaves and gently teasing it out.

Hold the seedling in one hand. Make a hole with the dibber in fresh compost in a new container.

Seedlings in shallow seedtrays need transplanting promptly as the roots have less room to develop in. Most young seedlings should be transplanted 2.5–4cm (1–1½in) apart in a seedtray or to a 9cm (3½in) pot, but slow growers may do better if moved first to a 5–8cm (2–3in) pot and later planted in a larger pot once they have a good root system.

Potting compost contains enough food for about five or six weeks. After this time plants will need supplementary feeding. Feed pricked-out seedlings and plug plants fortnightly with a balanced liquid fertiliser. Larger plants may need weekly applications but tiny seedlings may require only monthly feeding.

GERMINATING SEEDS

As a seed germinates, the embryo develops an initial root system below ground and a shoot system above. The seed leaves (cotyledons) may remain under the soil surface or emerge above with the shoot. They are usually simply shaped and generally do not look like the plant's true, mature leaves.

If seedlings become lanky, pinch out their shoot tips to encourage branching; this can be repeated several times.

Place the seedling roots in the hole and firm the compost gently around the roots with the dibber.

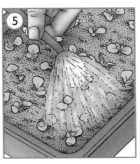
Water in the seedlings. Place the new container in a warm, well-lit area out of direct sunlight.

Harden off in a cold frame, raising the lid on warm days to acclimatise the seedlings to outdoor conditions.

• SEED SOWING TROUBLESHOOTER •

PROBLEM	CAUSE	SOLUTION
Seeds do not germinate	Seeds not viable (dead)	Buy fresh seeds. Most seeds can be stored for no more than three years in a cool (max. 10°C/50°F), dry place.
	Compost too dry or too wet	Sow in moist compost. Water and drain after sowing. Water with a sprayer only if the compost is dry below the surface.
	Wrong temperature	Seeds can rot if too cold or cook if too hot. Check the temperature needed for your seeds (for example, on the seed packet) and use a max.–min. thermometer to monitor the ambient temperature in the germination area.
	Seeds require special treatment	Unusual seeds may need different treatment; check the seed packet for special requirements.
Seeds germinate but do not emerge	Capping – the seedlings are unable to push though a crust of dried compost	Do not overheat the container and check seedlings daily. If the dry surface of the compost is heaving, use a sprayer to rewet it so the seedlings can push through.
	Seeds sown too deep	Use fine vermiculite to cover very fine seeds – they often need light to germinate and vermiculite lets it through. Also it is easier to cover the seeds thinly.
Seedlings emerge in patches	Uneven conditions in compost	Firm and level compost carefully before sowing and water the compost from below.
	Uneven sowing	Mix small seeds with fine vermiculite or sand so that you can see where you have just sown them.
	Seeds naturally variable	Some seeds, such as those of wild flowers, germinate erratically. Home-saved seeds may also be variable.

PROBLEM	CAUSE	SOLUTION
Seedlings are etoliated (thin, pale and drawn)	Too much heat and not enough light	If you cannot provide more light, keep seedlings at a lower temperature.
Seedlings are lopsided	Uneven light	Rotate the container daily so the seedlings are evenly exposed.
Seedlings wither	Not enough moisture	Check daily and water just enough to keep the compost moist.
	Too much heat, often from direct sun, can wither leaves even if the compost is moist	Maintain a humid atmosphere around newly emerged seedlings and shade them from direct sunlight.
Seedlings collapse	Seedlings affected by damping off (which rots the stems near the base)	This can occur soon after germination, or after pricking out, and is caused by a fungal infection, so be scrupulous about hygiene (see p33).
Seedlings fail to grow on	Too little warmth	Newly germinated seedlings require similar temperatures to those for germination. Harden them off slowly to avoid a check in growth.
	Too few nutrients	Seed composts contain only enough nutrients for germination: prick out seedlings promptly or liquid feed.
	Too much water	Do not overwater small seedlings in large pots or newly pricked-out seedlings until they are established.

GRAFTING

• INTRODUCTION TO GRAFTING •

Grafting is a skill that can take a while to master but is well worth doing. For some trees, such as growing fruit trees on dwarfing rootstocks, grafting is essential.

The purpose of grafting is to combine the ornamental or fruiting qualities of one plant (the scion) with the roots of another that offers vigour, resilience and sometimes disease resistance (the rootstock). Grafting exposes the cambial layer (the green regenerative layer just below the bark) of each plant and firmly binds them together. The wound forms a callus, and the scion and rootstock bond to form the new plant at a point known as the graft union.

High-value ornamental shrubs, such as acer, witch hazel (*Hamamelis*) and magnolia, are generally grafted while growing in a container. Because they are grafted under cover, this is known as bench grafting. Trees and roses are grafted onto rootstocks growing outdoors in the ground, and this is known as field grafting.

Fruit trees tend to be grafted for the following reasons:

● They may be too vigorous or exhibit unreliable growth characteristics when grown on their own root system.

● Grafting can invigorate a weak-growing cultivar.

● A fruiting plant can be produced in a shorter period of time than by other propagation methods.

● Cultivars will not usually come true from seed.

Some ornamental shrubs and trees are grafted for the following reasons:

● They are difficult to propagate by other means, such as cuttings.

● To strengthen plants that grow weakly on their own root systems.

● To produce a larger flowering plant in a shorter period of time.

Before you start grafting, you need to choose a suitable rootstock. These can be bought from rootstock growers and nurseries that specialise in the type of plant in question. Alternatively, they can be raised from seed or cuttings.

Choose a rootstock with desirable characteristics, such as a dwarfing habit or disease resistance, or a rootstock that is easier to propagate than the scion.

Most plants need to be grafted within their own species, such as *Acer palmatum* cultivars onto an *A. palmatum* rootstock. It is often possible to graft within a genus: *A. japonicum* can also be grafted onto *A. palmatum*. A few plants can be grafted onto different species, providing they are in the same family. For example, *Fothergilla* onto *Parrotia* rootstock, as both are in the *Hamamelidaceae*.

When grafting pine trees, a three-needled pine should be grafted onto a three-needled pine, and the same applies for five- and two-needled ones.

Use healthy material and a very sharp knife that is regularly sterilised and cuts straight through the wood so surfaces meet flush.

You will also need grafting tape and/ or grafting wax. These are available from specialist suppliers. However, you can make your own grafting tape from strips cut from a good-quality plastic bag. Some grafters use elasticated bands instead or even raffia.

Apples and other fruit trees are grafted onto a rootstock to control their height.

WHIP & TONGUE GRAFTING

This method is mainly used to propagate fruit trees, but can also be adopted for ornamental trees and shrubs, which are grafted as close to the ground as possible to help prevent the graft union being obvious. It is normally carried out in spring on a rootstock that was planted 12 months previously. This technique uses two cuts on both the scion and rootstock, which enables the two parts to be locked together with a 'tongue', so giving a structurally strong graft suitable for field conditions.

Failure can happen when the callus between the scion and rootstock does not form. This can be due to poor cutting and joining, or inferior-quality plant material.

PROPAGATING BY WHIP & TONGUE GRAFTING

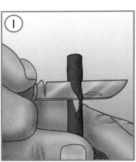

1. Just before bud break, cut back the rootstock where the scion is to be grafted and trim off all sideshoots. Fruit trees are usually grafted 23–30cm (9–12in) above soil level. Make a 4cm (1½in) upward-sloping cut at the top.

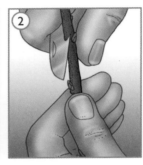

2. Select a suitable, healthy scion and make a top cut just above a bud that is about four buds from the base.

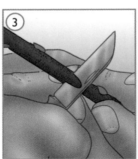

3. Make a 4cm (1½in) basal cut at the same angle as that on the rootstock. End it just below the bottom bud.

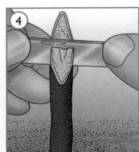

4. Make a shallow, single 3–6mm (⅛–¼in) slice into the rootstock from one third down the sloping cut.

Sometimes graft unions can fail many years after they were, seemingly, successfully made. This is commonly seen in wisteria, where the reason given is graft incompatibility.

In winter, select healthy and vigorous, one-year-old shoots from the scion that are suitable as hardwood stems. Both the rootstock and scion material need to be about the same diameter – preferably 2.5cm (1in). Remove a 23–30cm (9–12in) length by cutting just above a bud on the parent plant.

Bundle five or six scions together and heel them into a well-drained, sheltered site, leaving 5–8cm (2–3in) showing above the soil. This will keep the scions moist but dormant. Alternatively, wrap them up in a dry plastic bag and keep them in the fridge until spring.

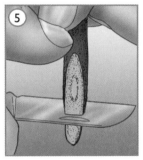

Make a matching 3–6mm (⅛–¼in) slice into the scion from one third up its sloping cut.

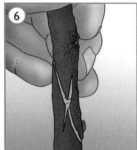

Slip the scion into the rootstock so they interlock. Ensure the cambiums (see p136) in the two halves match up as well as possible.

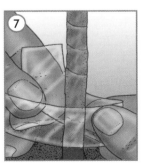

Bind the join firmly with clear polythene tape or grafting tape. Dab the top cut end of the scion with tree paint.

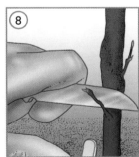

Once the cut surfaces start callusing, remove the polythene tape. This should have taken about eight weeks. Cut off any rootstock growth. For a single-stemmed tree remove all but one of the scion shoots.

• APICAL-WEDGE GRAFTING •

Also known as cleft grafting, this method is used to propagate a number of ornamental trees and shrubs, such as hibiscus (pictured), syringa and wisteria.

Wedge grafting is the easiest way of joining two stems and probably one of the first methods to attempt. Although the cuts in the rootstock and the scion need to line up to callus properly, it is not so exacting and critical as some other grafting methods. The vertical cut in the rootstock holds a wedge of the scion firmly in place, so the success rate is usually quite high.

In midwinter, select healthy and vigorous, one-year-old shoots from the scion that are suitable as hardwood stems. Remove a suitable length by cutting just above a bud on the parent

Named cultivars of Hibiscus are propagated by apical wedge grafting.

PROPAGATING BY APICAL-WEDGE GRAFTING

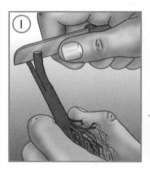 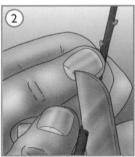 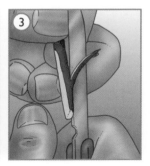

Make a single 3cm (1¼in) cut vertically down the centre of the chosen rootstock.

Make a sloping top cut on the scion just above a bud and a horizontal one about 15cm (6in) below it.

Make a 4cm (1½in) cut towards the middle of the scion base. Repeat on the other side to form a 'V'.

plant. The rootstock and scion material should be about the same diameter.

Bundle five or six scions together and heel them into a well-drained, sheltered site in the garden, leaving 5–8cm (2–3in) showing above the soil. This will keep them moist but dormant.

In late winter or early spring, select a suitable rootstock, lift it and wash the soil from the roots. Cut the stem horizontally across, approximately 5–8cm (2–3in) above the roots, then cut vertically down the centre of the stem.

The small portion of the cut scion surface exposed above the cut in the rootstock when they are taped together will help encourage the formation of callus tissue; this exposed area is known as the 'church window'.

Heel the plants into a box or pot filled with multipurpose compost, inserting them so the compost is just above the graft union. Place the container in a cold frame or propagator at 10–15°C (50–59°F). The higher the temperature, the quicker the union will happen, providing the plants do not dry out.

As the grafted plants unite, the exposed cut surfaces will start to callus and this can be seen in the church window.

The tape can be removed when the union is firm. The newly formed plant can be potted up before being gradually hardened off for planting out.

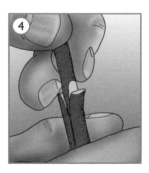
Push the scion into the rootstock cut. Leave a small part of the scion cut surface exposed.

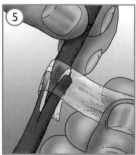
Thoroughly bind the joined area together with clear polythene tape or grafting tape.

Place the plants in a box filled with compost to cover the graft union. Place the box in a protected area.

• SIDE-WEDGE GRAFTING •

The side-wedge, or side-splice, graft is the easiest and most successful method of side grafting. It can be used for a wide range of both deciduous and evergreen plants, including rhododendrons, birch (*Betula*) and witch hazel (*Hamamelis*).

Grafting is carried out in late winter or early spring, just before growth starts.

PROPAGATING BY SIDE-WEDGE GRAFTING

Make two sloping cuts, 4cm (1½in) long and opposite each other, at the base of the scion.

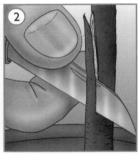

From 5cm (2in) above the base of the rootstock, cut down and inwards to make a cut 4cm (1½in) long.

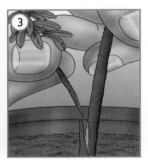

Bend the rootstock gently away from this cut. Hold it there and insert the scion. Then release it.

Bind the entire cut with clear polythene tape. Keep the grafted rootstock in a warm glasshouse.

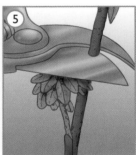

Once the two parts have joined, remove the tape and cut back half the length of the rootstock.

Two weeks later cut back the remaining rootstock above the grafted union.

Deciduous azalea plants are produced by side-wedge grafting.

Evergreens can also be grafted whenever a mature flush of growth is available and the rootstock is actively growing.

For very sappy plants, especially those that bleed when cut, graft early in the growing season and dry off the rootstocks before using them.

Plant the rootstock in a pot of good-quality compost and grow on for a year so it is well established. About three or four weeks before grafting is to be done, place the rootstock in a cool, frost-free area to encourage it into growth. Keep it dry by watering only minimally or not at all. Just before bud break remove the leaves from the bottom 8–10cm (3–4in) of the rootstock stem.

Select suitable scion material, preferably with the apical bud intact, and remove from the parent plant with all the previous season's growth. Make two sloping cuts at the base to produce a wedge shape.

Cut downwards into the rootstock, starting about 5cm (2in) above ground level and slightly slanting the cut inwards to end one third of the way through the stem.

Keep the grafted plant in a warm glasshouse. The two parts should join within six to eight weeks.

Treat the cut end on the rootstock with a wound paint or warm sealing wax if it bleeds at all after it has been shortened back to just above the scion, which becomes the leader shoot of the plant.

Water little and often rather than in large doses at infrequent intervals, so the plant is kept dry in a humid setting.

• CHIP-BUDDING •

In this propagation technique a bud on a sliver of wood, complete with the bark – the chip – is inserted into a matching notch on the rootstock. Chip-budding is used for increasing deciduous fruit and ornamental trees. Trees in the rose family, such as apple, cherry, hawthorn, pear, plum and sorbus, are especially suitable for this method.

Chip-budding can be done at most times of the year, especially between summer and early autumn, when the temperature is above 10°C (50°F) and well-matured buds are available.

Failure of the buds to take is usually as a result of not cutting the bud chip and rootstock accurately enough to get the cambium layers (see p136) to match precisely. It pays to practise on spare shoots until you can reliably cut a really good match. Some less experienced gardeners attach several buds as at least one should take.

Knowing when to remove the bud ties can be difficult and is something that comes with experience. The bud and the cambium must have united, and this is indicated by swelling of the budded part of the stem.

From midsummer choose the buds from the scion you want to use by selecting non-flowering shoots that are

PROPAGATING BY CHIP-BUDDING

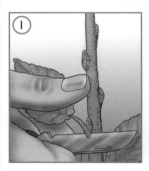

Remove the softer topgrowth from the 'budstick'. Cut away all the leaves flush with the stem.

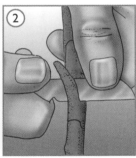

Make a cut 6mm (¼in) deep in the rootstock. To remove a chip cut down to it from 4cm (1½in) up the stem.

Make exactly similar cuts on the budstick, ensuring that a bud is included midway down the chip.

a similar diameter to the rootstock, from well-ripened, current year's growth. Remove these 'budsticks' from the parent plant, so that they are ready to be budded onto the rootstock.

To prepare the rootstock, cut off all shoots and leaves from the bottom 30–45cm (12–18in) of the stem. Using a clean, very sharp knife make a downward cut 2cm (¾in) below a bud, at an angle of 30–45 degrees. Make a second cut about 4cm (1½in) above the first. Cut down through the wood to meet the first cut, taking care not to damage the bud.

To prepare the budstick, discard the topgrowth and foliage. Then immediately make two cuts in the rootstock about 15cm (6in) from the ground to correspond exactly with those on the rootstock and remove the resulting sliver of wood – the bud chip.

Place the bud chip into the 'lip' of the cut rootstock so that the cambium layers match as well as possible. Bind the join with grafting tape or strips of polythene so that it completely seals the chip.

After three or four weeks the tape can be removed once the bud starts to swell. Insert a cane and tie in the new shoot as it develops.

The following late winter or spring, cut back the rootstock just above the bud. Plant out the following winter after the bud has grown into a new tree.

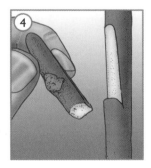
Tuck the bud chip into position on the rootstock so they match. Cover the chip with clear polythene tape.

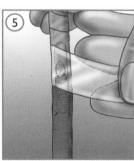
Once the bud has united with the rootstock and starts to swell remove the tape.

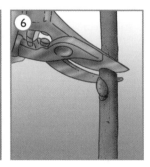
In winter cut back the rootstock, close above the bud. The bud will grow out the next season.

PROPAGATION DIRECTORY

ANNUALS, BIENNIALS & BEDDING PLANTS
FOR FLOWERS AND FOLIAGE – *AGERATUM* TO *CENTAUREA*

Plant name	Plant type	Propagation method
Ageratum houstonianum (floss flower)	Half-hardy summer bedding plant	Sow seed (see p126) at 16–18°C (61–64°F) in early spring
Amaranthus cordatus (love-lies-bleeding)	Hardy annual	Sow seed in containers (see p126) at 20°C (68°F) or in situ (see p116) in midspring
Anchusa capensis (alkanet)	Hardy annual	Sow seed in containers (see p126) at 13–16°C (55–61°F) in late winter or early spring
Antirrhinum majus (snapdragon)	Half-hardy summer bedding plant	Sow seed in containers (see p126) at 16–18°C (61–64°F) in late summer, early winter or early spring and overwinter young plants under glass
Arctotis (African daisy)	Half-hardy summer bedding plants	Sow seed in containers (see p126) at 16–18°C (61–64°F) in early spring and prick out into small pots or modules to reduce root disturbance; or take softwood cuttings (see p44) at any time
Begonia Semperflorens Group (fibrous-rooted begonia)	Half-hardy summer bedding plants	Sow seed in containers (see p126) at 21°C (70°F) in early spring; or root softwood cuttings (see p44) in summer
Bellis perennis (double daisy)	Hardy spring bedding plant	Sow seed (see p126) at 10–13°C (50–55°F) in early spring or in shallow drills outdoors (see p116) in early spring; or divide (see p98) in early spring or after flowering
Bidens ferulifolia	Half-hardy summer bedding plant	Sow seed (see p126) at 13–18°C (55–64°F) in spring; or root softwood cuttings (see p44) in spring or autumn
Brachyscome iberidifolia (swan river daisy)	Half-hardy summer bedding plant	Sow seed in containers (see p126) at 18°C (64°F) in spring
Calendula officinalis (pot marigold)	Hardy annual	Sow seed in situ (see p116) in spring
Calibrachoa (trailing petunia)	Half-hardy summer bedding plants	Take softwood cuttings (see p44) in late spring or summer
Callistephus chinensis (Chinese aster)	Half-hardy summer bedding plant	Sow seed in containers (see p126) at 16°C (61°F) in early spring or in situ (see p116) in midspring
Campanula medium (Canterbury bells)	Hardy biennial border plant	Sow seed in containers (see p126) in a cold frame in spring
Centaurea cyanus (cornflower)	Hardy annual	Sow seed in situ (see p116) in midspring or early autumn

ANNUALS, BIENNIALS & BEDDING PLANTS
FOR FLOWERS AND FOLIAGE – *CERINTHE* TO *ESCHSCHOLZIA*

Plant name	Plant type	Propagation method
Cerinthe major 'Purpurascens'	Half-hardy border plant	Sow seed in containers (see p126) at 13–16°C (55–61°F) in spring
Clarkia amoena (godetia, satin flower)	Hardy annual	Sow seed in situ (see p116) in early spring
Cleome hassleriana (spider flower)	Half-hardy summer bedding plant	Sow seed in containers (see p126) at 18°C (64°F) in spring
Cobaea scandens (cup-and-saucer plant)	Half-hardy perennial climber	Sow seed in containers (see p126) at 18°C (64°F) in spring; or root softwood cuttings (see p44) with bottom heat in summer
Consolida ajacis (larkspur)	Hardy annual	Sow seed in situ (see p116) from early spring to early summer
Convolvulus tricolor	Hardy annual	Sow seed in situ (see p116) in midspring
Coreopsis tinctoria (tickseed)	Hardy annual	Sow seed in situ (see p116) from early spring to early summer
Cosmos bipinnatus	Half-hardy summer bedding plant	Sow seed in containers (see p126) at 16°F (61°C) in midspring or in situ (see p116) in late spring
Dahlia, bedding varieties	Half-hardy summer bedding plants	Sow seed in containers (see p126) at 16°C (61°C) in early spring. (For varieties grown from tubers see p156.)
Dianthus barbatus (sweet William)	Hardy spring bedding plant	Sow seed in situ (see p116) in autumn or at 13°C (55°F) (see p126) in early spring to flower the same year
Dianthus chinensis (Indian pink)	Half-hardy summer bedding plant	Sow seed in containers (see p126) at 13–15°C (55–59°F) in early spring
Digitalis purpurea (foxglove)	Hardy biennial border plant	Sow seed in containers (see p126) in a cold frame in late spring or in situ (see p116) in late spring.
Eccremocarpus scaber (Chilean glory flower)	Hardy annual climber	Sow seed in containers (see p126) at 13–16°C (55–61°F) in late winter or early spring; or root softwood cuttings (see p44) with bottom heat in spring or summer
Echium vulgare, dwarf varieties	Hardy annuals	Sow seed in situ (see p116) or in containers (see p126) in spring
Erysimum cheiri (wallflower)	Hardy spring bedding plant	Sow seed in a seedbed (see p116) from late spring to early summer, grow on in a nursery bed and move to final positions in midautumn
Eschscholzia californica (California poppy)	Hardy annual	Sow seed in situ (see p116) in midspring or early autumn

ANNUALS, BIENNIALS & BEDDING PLANTS
FOR FLOWERS AND FOLIAGE – *GAZANIA* TO *MYOSOTIS*

Plant name	Plant type	Propagation method
Gazania	Half-hardy summer bedding plants	Sow seed (see p126) at 18–20°C (64–68°F) in late winter or early spring; or take basal cuttings (see p44) in late summer or early autumn and overwinter under glass
Helianthus annuus (sunflower)	Hardy annual	Sow annuals at 16°F (61°C) in containers (see p126) in late winter or in situ (see p116) in spring
Heliotropium arborescens (heliotrope)	Half-hardy summer bedding plant	Sow seed (see p126) at 16–18°C (61–64°F) in spring; or take softwood (see p44) or semiripe (see p48) cuttings in summer and overwinter in a frost-free place
Iberis umbellata (candytuft)	Hardy annual	Sow in situ (see p116) in spring or autumn
Impatiens walleriana (busy Lizzie)	Tender perennial bedding plant	Sow seed (see p126) at 16–18°C (61–64°F) in early spring; or root softwood cuttings (see p44) in spring or early summer
Ipomoea, annual species (morning glory)	Half-hardy annual climbers	Soak or chip seed (see p122) then sow singly in containers (see p126) at 18°C (64°F) in spring
Isotoma axillaris	Tender perennial summer bedding plant	Sow seed in containers (see p126) at 16–18°C (61–64°F) in spring; or root softwood cuttings (see p44) in summer and overwinter in a frost-free place
Lathyrus odoratus (sweet pea)	Hardy annual climber	Soak or chip seed (see p123), then sow in containers (see p126) in a cold frame in early spring or autumn or in situ (see p116) in midspring
Lavatera trimestris (mallow)	Hardy annual	Sow (see p126) under glass in midspring or in situ (see p116) from mid- to late spring
Limnanthes douglasii (poached-egg plant)	Hardy annual	Sow seed in situ (see p116) in spring
Lobelia erinus	Half-hardy summer bedding plant	Sow seed in containers (see p126) at 13–18°C (55–64°F) in late winter
Lobularia maritima (sweet alyssum)	Hardy annual	Sow seed in situ (see p116) in late spring
Lunaria annua (honesty)	Hardy biennial border plant	Sow seed in a seedbed (see p116) in early summer
Matthiola incana (stock)	Hardy spring or summer bedding plant	For spring bedding sow in containers (see p126) in a cold frame in midsummer, overwinter under glass and plant out in spring. For summer bedding sow in containers (see p126) at 10–18°F (50–64°C) in early spring
Myosotis sylvatica (forget-me-not)	Hardy spring bedding plant	Sow seed in containers (see p126) in a cold frame or seedbed (see p116) in early summer

ANNUALS, BIENNIALS & BEDDING PLANTS
FOR FLOWERS AND FOLIAGE – *NEMESIA* TO *PETUNIA*

Plant name	Plant type	Propagation method
Nemesia strumosa	Half-hardy summer bedding plant	Sow seed in containers (see p126) at 15°C (59°F) in spring
Nemophila menziesii (baby blue-eyes)	Hardy annual	Sow seed in situ (see p116) in early spring or autumn
Nicandra physalodes (shoo fly)	Hardy annual	Sow seed in situ (see p116) in midspring or in containers (see p126) at 15°C (59°F) in early spring
Nicotiana alata	Half-hardy summer bedding plant	Surface sow seed in containers (see p126) at 18°C (64°F) in midspring
Nigella damascena (love-in-a-mist)	Hardy annual	Sow seed in situ (see p116) in midspring
Oenothera biennis (evening primrose)	Hardy biennial border plant	Sow seed in containers (see p126) in a cold frame in early summer or in situ (see p116) in autumn
Papaver croceum (Icelandic poppy)	Hardy biennial border plant	Sow seed in situ (see p116) in spring
Papaver somniferum (opium poppy)	Hardy annual	Sow seed in situ (see p116) in spring
Pelargonium, zonal varieties (geranium)	Tender perennial summer bedding plant	Sow seed in containers (see p126) at 13–18°C (55–64°F) in late winter and early spring; or take softwood cuttings (see p44) in spring, late summer or early autumn.
Petunia	Half-hardy summer bedding plants	Sow seed in containers (see p126) at 13–18°C (55–64°F) in midspring; or take softwood cuttings (see p44) in summer and overwinter young plants under glass

Pot marigold (Calendula officinalis) *is a hardy annual; sow seeds direct in spring.*

Cosmos bipinnatus *can be sown indoors with heat or outside in situ.*

Zinnia seed is best germinated with warmth in a heated propagator.

ANNUALS, BIENNIALS & BEDDING PLANTS
FOR FLOWERS AND FOLIAGE – *PHLOX* TO *ZINNIA*

Plant name	Plant type	Propagation method
Phlox drummondii (annual phlox)	Half-hardy summer bedding plant	Sow seed in containers (see p126) at 13–18°C (55–64°F) in early spring
Primula Polyanthus Group	Hardy spring bedding plants	Sow seed in containers (see p126) in a cold frame in summer and plant out in autumn; or divide plants (see p98) after flowering
Rudbeckia hirta (black-eyed Susan)	Hardy annual and biennial border plant	Sow seed in containers (see p126) at 16–18°C (61–64°F) in spring
Salvia farinacea (mealy sage)	Half-hardy summer bedding plant	Sow seed in containers (see p126) at 16–18°C (61–64°F) in midspring
Salvia splendens (scarlet sage)	Half-hardy summer bedding plant	Sow seed in containers (see p126) at 16–18°C (61–64°F) in midspring
Scaevola aemula (fairy fan-flower)	Tender perennial trailer	Sow seed in containers (see p126) at 19–24°C (66–75°F) in spring; or root softwood cuttings (see p44) with bottom heat in late spring or summer
Senecio cineraria (dusty miller)	Half-hardy summer bedding plant	Sow seed (see p126) at 19–24°C (66–75°F) in spring; or take semiripe cuttings (see p48) in mid- or late summer and overwinter in a frost-free place
Tagetes (African and French marigolds)	Half-hardy summer bedding plants	Sow in containers (see p126) at 21°C (70°F) in early spring or in situ (see p116) in late spring
Thunbergia alata (black-eyed Susan)	Tender perennial climber	Sow seed in containers (see p126) at 16–18°C (61–64°F) in spring
Tithonia rotundifolia (Mexican sunflower)	Half-hardy annual	Sow seed in containers (see p126) at 13–18°C (55–64°F) from mid- to late spring
Tropaeolum majus (nasturtium)	Half-hardy summer bedding plant	Sow seed in situ (see p116) in midspring or in containers (see p126) at 13–18°C (55–64°F) in early spring
Tropaeolum peregrinum (canary creeper)	Annual climber	Sow seed in situ (see p116) in midspring or in containers (see p126) at 13–18°C (55–64°F) in early spring
Verbena × *hybrida*	Half-hardy summer bedding plant	Sow seed in containers (see p126) at 18–21°C (64–70°F) in early spring
Viola × *wittrockiana* (pansy)	Hardy spring or summer bedding plant	Sow seed in containers (see p126) in a cold frame in late winter for spring and summer flowering and in summer for winter flowering
Xerochrysum bracteatum (strawflower)	Hardy annual	Sow seed in containers (see p126) at 18°C (64°F) in spring
Zinnia elegans	Half-hardy summer bedding plant	Sow seed in containers at 13–18°C (55–64°F) in early spring (see p126) or in situ (see p116) in late spring

PERENNIALS FOR FLOWERS AND FOLIAGE – *ACAENA* TO *ANEMONE*

Plant name	Plant type	Propagation method
Acaena (New Zealand burr)	Evergreen dwarf perennials	*Separate rooted stems (see p98) in autumn or early spring; or take softwood cuttings (see p44) in late spring*
Acanthus spinosus (bear's breeches)	Border perennial	*Take root cuttings (see p74) in winter; or sow seeds as soon as ripe in containers (see p126) in a cold frame and grow on for two years before planting out*
Achillea (yarrow)	Border perennials	*Divide (see p98) in spring; or sow seed in situ (see p116) in spring*
Aconitum (monkshood)	Border perennials	*Divide (see p98) in autumn or late winter; or sow seed in containers (see p126) in a cold frame in spring*
Agapanthus (African lily)	Border perennials	*Divide (see p98) in spring; or sow seed as soon as ripe or in spring in containers (see p126) in a cold frame and plant out the following spring*
Agastache	Border perennials	*Divide (see p98) in spring; take semiripe cuttings (see p48) in late summer and overwinter under glass; or sow seed in containers (see p126) at 13–18°C (55–64°F) in early spring*
Agave	Tender perennial succulents	*Remove offsets (see p100) in spring or autumn and plant out if rooted, or pot up in gritty compost until established*
Ajuga (bugle)	Border perennials	*Separate rooted stems (see p98); or take softwood cuttings (see p44) in early summer*
Alcea (hollyhock)	Border perennials	*Sow seed in situ (see p116) in midsummer – young plants can be moved in autumn if necessary*
Alchemilla (lady's mantle)	Border perennials	*Divide (see p98) in early spring or autumn; or sow seed in containers (see p126) in a cold frame in spring and plant seedlings out while still small*
Allium, ornamental species (ornamental onion)	Perennials grown from bulbs	*Divide clumps (see p98) in autumn; separate bulblets (see p86) in autumn; or sow seed in containers (see p126) in a cold frame as soon as ripe or in spring. Some species produce bulbils, which can be planted in containers in a cold frame or in a seedbed when ripe (see p86)*
Alstroemeria (Peruvian lily)	Border perennials	*Divide established clumps (see p98) in autumn or very early spring*
Amaryllis	Perennials grown from bulbs	*Separate bulblets (see p86) in spring and grow them on under glass for a year or two before planting out*
Anchusa (alkanet)	Border perennials	*Take basal softwood cuttings (see p44) in spring; or take root cuttings (see p74) in winter*
Anemone blanda, *A. coronaria*	Perennials grown from tubers	*Separate clumps of tubers (see p82) in summer, when dormant*

PERENNIALS FOR FLOWERS AND FOLIAGE – *ANEMONE* TO *ASTER*

Plant name	Plant type	Propagation method
Anemone hupehensis (Japanese anemone)	Border perennial	Divide (see p98) in early spring or autumn and grow on in containers for 12–18 months before planting out in spring
Anthemis	Border perennials	Divide (see p98) in spring; or take basal softwood cuttings (see p44) in spring or late summer
Anthericum (St Bernard's lily)	Border perennials	Divide (see p98) as growth begins in spring
Aquilegia (columbine)	Border perennials	Sow seed in containers (see p126) in a cold frame as soon as ripe or in spring. Divide named cultivars (see p98) in spring
Arabis (rock cress)	Dwarf evergreen perennials	Root softwood cuttings (see p44) in summer; or sow seed in containers (see p126) in a cold frame in autumn
Armeria (sea pink, thrift)	Dwarf evergreen perennials	Divide (see p98) in early spring; take semiripe basal cuttings (see p48) in summer; or sow seed in containers (see p126) in a cold frame in spring or autumn
Artemisia	Border perennials	Divide (see p98) in spring or autumn; or take greenwood cuttings (see p47) or heel cuttings (see p56) of sideshoots in early summer
Arum (lords-and-ladies)	Perennials grown from tubers	Divide clumps of tubers (see p82) after flowering
Aster (Michaelmas daisy)	Border perennials	Divide (see p98) in spring. Take basal softwood cuttings (see p44) of A. amellus, A. × frikartii and A. thomsonii in spring

Agaves are easy to propagate by removing offsets from mature plants.

Divide cannas in early spring, each section having at least one bud.

Dahlias can be grown from softwood cuttings or by dividing tubers.

PERENNIALS FOR FLOWERS AND FOLIAGE – *ASTILBE* TO *CHIONODOXA*

Plant name	Plant type	Propagation method
Astilbe	Border perennials	Divide (see p98) in winter or early spring
Astrantia (masterwort)	Border perennials	Divide (see p98) in spring; or sow seed in containers (see p126) in a cold frame as soon as ripe
Aubrieta × *cultorum*	Dwarf perennial	Sow seed in containers (see p126) in a cold frame in autumn or spring; take softwood cuttings (see p44) in early summer or semiripe cuttings (see p48) in midsummer
Aurinia saxatilis (gold dust)	Dwarf evergreen perennial	Take greenwood cuttings (see p47) in early summer; or sow seed in containers (see p126) in a cold frame in autumn or spring
Bergenia (elephant's ears)	Evergreen border perennials	Divide clumps (see p98) in autumn or spring; or root sections of young rhizome (see p94), with one or more leaf buds, in autumn
Brunnera macrophylla	Border perennial	Divide (see p98) in spring; or take root cuttings (see p74) in winter
Camassia (quamash)	Perennials grown from bulbs	Separate bulblets (see p86) when dormant in summer; or sow seed as soon as ripe in containers (see p126) in a cold frame
Campanula, taller species (bellflower)	Border perennials	Divide (see p98) in spring or autumn; take basal softwood cuttings (see p44) in spring; or sow seed in containers (see p126) in a cold frame in spring
Canna (Indian shot)	Tender perennials	Divide rhizomes (see p94) into short sections, each with an eye, in early spring, pot up and start into growth at 16°F (61°C); or chip or soak seed in warm water for 24 hours (see p122) before sowing (see p126) at 21°C (70°F)
Carex	Evergreen ornamental grasses	Divide (see p98) from midspring to early summer. Sow hardy species in autumn in containers in a cold frame and half-hardy species at 10–13°C (50–55°F) in containers (see p123) in early spring
Centaurea, perennial species (knapweed)	Border perennials	Divide (see p98) in spring or autumn. Take root cuttings (see p74) of C. montana *in winter*
Centranthus ruber (red valerian)	Border perennial	Sow seed in situ (see p116) in autumn or spring
Cephalaria gigantea (giant scabious)	Very tall border perennial	Divide (see p98) in early or midspring; or sow seed in containers (see p126) in a cold frame in early spring
Chionodoxa (glory of the snow)	Perennials grown from bulbs	Separate bulblets (see p86) in summer; sow seed in containers (see p126) in a cold frame as soon as ripe; or use bulb scales (see p87) or bulb scoring (see p89)

PERENNIALS FOR FLOWERS AND FOLIAGE – *CIMICIFUGA* TO *DAHLIA*

Plant name	Plant type	Propagation method
Cimicifuga (bugbane)	Border perennials	Divide (see p98) in spring; or sow seed in containers (see p126) in a cold frame as soon as ripe
Colchicum (autumn crocus)	Perennials grown from corms	Separate corms (see p90) in summer; or sow seed in containers (see p126) in a cold frame as soon as ripe
Convallaria (lily of the valley)	Border perennials	Divide (see p98) in autumn by separating rhizomes
Coreopsis (tickseed)	Border perennials	Sow seed in a seedbed (see p116) in midspring or in containers (see p126) at 13–16°C (55–61°F) in late winter; or divide (see p98) in early spring
Cortaderia (pampas grass)	Ornamental perennial grasses	Divide young plants (see p98) in spring; or sow seed in containers (see p126) at 13–18°C (55–64°F) in spring
Corydalis (fumitory)	Dwarf perennials	Divide spring-flowering species (see p98) in autumn and summer-flowering species in spring
Cosmos atrosanguineus (chocolate cosmos)	Border perennial	Take basal softwood cuttings (see p44) with bottom heat in early spring
Crambe cordifolia	Very tall border perennial	Sow seed in containers (see p126) in a cold frame in spring or autumn; take root cuttings (see p74) in autumn; or divide (see p98) in early spring
Crinum × *powellii*	Perennial grown from bulbs	Remove offsets (see p100) in spring
Crocosmia (montbretia)	Perennials grown from corms	Divide (see p98) in spring, just before growth starts; or separate cormels (see p91) in autumn
Crocus	Dwarf perennials grown from corms	Separate cormels (see p91) when dormant; or sow seed in containers (see p126) in a cold frame as soon as ripe
Cyclamen, hardy species	Dwarf perennials grown from tubers	Soak seed as soon as ripe for 24 hours and rinse, then sow in containers (see p126) at 6–12°C (43–54°F)
Cynara cardunculus (cardoon)	Border perennial	Sow seed in containers (see p126) in a cold frame in spring; divide (see p98) in spring; or take root cuttings (see p74) in winter
Dactylorhiza (marsh orchid, spotted orchid)	Perennials grown from tubers	Divide (see p98) in early spring
Dahlia	Half-hardy perennials grown from tubers	Start tubers into growth under glass in late winter or early spring, then take basal softwood cuttings (see p44); or divide tubers (see p82), ensuring each division has at least one shoot. (For bedding varieties see p149.)

PERENNIALS FOR FLOWERS AND FOLIAGE – *DELPHINIUM* TO *EREMURUS*

Plant name	Plant type	Propagation method
Delphinium	Border perennials	For varieties in Elatum and Belladonna Groups take basal softwood cuttings (see p44) with solid heels in early spring; or divide (see p98) in spring. For other Delphinium types sow seed (see p126) at 13°C (55°F) in early spring
Dianthus (border carnations and pinks)	Border perennials	Take cuttings (see p44) from non-flowering shoots in summer; or layer (see p104) after flowering
Diascia	Border perennials	Sow seed in containers (see p126) at 16°F (61°C) in early spring; or take semiripe cuttings (see p48) in summer and overwinter under glass
Dicentra (bleeding heart)	Border perennials	Divide (see p98) in early spring, or after the leaves have died down
Dictamnus albus (burning bush)	Border perennial	Sow seed in containers (see p126) in a cold frame as soon as ripe; or divide (see p98) in autumn or spring
Dierama (wand flower)	Perennials grown from corms	Divide in spring (see p98); separate cormels (see p91) in spring and grow on in a seedbed; or sow seed in containers (see p126) in a cold frame as soon as ripe
Digitalis, perennial species (foxglove)	Border perennials	Sow seed in containers (see p126) in a cold frame in late spring
Dodecatheon (shooting stars)	Border perennials	Divide (see p98) in spring; or sow seed in containers (see p126) in a cold frame as soon as ripe – seed needs exposure to cold to germinate
Doronicum (leopard's bane)	Border perennials	Divide (see p98) in early autumn; or sow seed in containers (see p126) in a cold frame in spring
Echinacea purpurea (coneflower)	Border perennial	Sow seed in containers (see p126) at 13°C (55°F) in spring; take root cuttings (see p74) from late autumn to early winter; or divide in autumn or spring (see p98)
Echinops ritro (globe thistle)	Border perennial	Divide (see p98) from autumn to spring; sow seed in a seedbed (see p116) as soon as ripe or in midspring; or take root cuttings (see p74) in winter
Epilobium (willow herb)	Border perennials	Divide (see p98) in autumn or spring; take softwood cuttings (see p44) from sideshoots in spring; or sow seed in containers (see p126) in a cold frame in spring
Epimedium (barrenwort)	Border perennials	Divide (see p98) in autumn or after flowering
Eranthis (winter aconite)	Dwarf perennials grown from tubers	Separate tubers (see p92) in spring after flowering
Eremurus (foxtail lily)	Border perennials	Divide (see p98) after flowering

PERENNIALS FOR FLOWERS AND FOLIAGE – *ERIGERON* TO *GERANIUM*

Plant name	Plant type	Propagation method
Erigeron (fleabane)	Border perennials	Divide (see p98) in spring; or take basal softwood cuttings (see p44) in spring
Erodium (stork's bill)	Border perennials	Divide (see p98) in spring; or take basal softwood cuttings (see p44) in late spring or early summer
Eryngium (sea holly)	Border perennials	Sow seed in containers (see p126) in a cold frame as soon as ripe or in spring; or take root cuttings (see p74) in late winter
Erysimum, perennial varieties (wallflower)	Evergreen border perennials	Take softwood cuttings (see p44) in spring or summer
Erythronium (dog's tooth violet)	Perennials grown from tubers	Divide (see p98) after flowering
Euphorbia, herbaceous species (spurge)	Border perennials	Take basal softwood cuttings (see p44) in spring or early summer and dip cut surface in warm water, or a flame, to prevent bleeding; or sow seed in containers (see p126) in a cold frame as soon as ripe or in spring
Ferula communis (giant fennel)	Very tall border perennial	Sow seed as soon as ripe, or in spring, in deep containers (see p126) in a cold frame
Festuca (fescue)	Ornamental evergreen grasses	Divide (see p98) in spring; or sow seed in containers (see p126) in a cold frame in autumn or spring
Filipendula (meadowsweet)	Border perennials	Divide (see p98) in autumn or spring; or sow seed in containers (see p126) in a cold frame in autumn or at 10–13°C (50–55°F) in spring
Fritillaria (fritillary)	Perennials grown from bulbs	Separate bulblets (see p86) in late summer
Gaillardia, perennial varieties (blanket flower)	Border perennials	Sow seed (see p126) at 13–18°C (55–64°F) in early spring; divide (see p98) in spring; take root cuttings (see p74) in winter; or take softwood cuttings (see p44) in summer
Galanthus (snowdrop)	Dwarf perennials grown from bulbs	Divide clumps of bulbs (see p86) as the leaves die down
Galtonia	Perennials grown from bulbs	Separate bulblets (see p86) in early spring
Gaura lindheimeri	Border perennial	Divide (see p98) in spring; take softwood cuttings (see p44) in spring; take semiripe heel cuttings (see p48) in summer; or sow seed in containers (see p126) in a cold frame in spring or early summer
Geranium, hardy species (cranesbill)	Border perennials	Divide (see p98) in spring; or sow seed in containers (see p126) in a cold frame as soon as ripe or in spring

PERENNIALS FOR FLOWERS AND FOLIAGE – *GEUM* TO *HEUCHERA*

Plant name	Plant type	Propagation method
Geum	Border perennials	Divide (see p98) in autumn or spring; or sow seed in containers (see p126) in a cold frame in autumn or spring
Gladiolus	Hardy and half-hardy perennials grown from corms	Separate cormels (see p91) in autumn; or use scoring (see p89). Sow seed of hardy species in containers (see p126) in a cold frame in spring
Gunnera manicata	Very tall border perennial	Take softwood cuttings (see p44) of leafy basal buds in spring
Gypsophila (baby's breath)	Border perennials	Sow seeds in containers (see p126) at 13–18°C (55–64°F) in late winter; or take basal softwood cuttings (see p44) from mid- to late spring
Hakonechloa macra	Ornamental grass	Divide (see p98) in spring
Helenium autumnale (sneezewort)	Border perennial	Divide (see p98) in autumn or spring; or root basal softwood cuttings (see p44) in spring
Helictotrichon sempervirens (blue oat grass)	Evergreen ornamental grass	Sow seed in containers (see p126) in a cold frame in spring; or divide (see p98) in spring
Helleborus (hellebore)	Border perennials	Divide young plants (see p98) after flowering, in early spring or in late summer; or sow seed in containers (see p126) in a cold frame as soon as ripe
Hemerocallis (daylily)	Border perennials	Divide (see p98) in spring or autumn
Heuchera sanguinea (coral bells)	Border perennial	Divide (see p98) in autumn. Sow seed of species in containers (see p126) in a cold frame in spring

Grow coneflowers (Echinacea) from seed, root cuttings or by dividing plants.

Divide the rhizomes of bearded irises after flowering or in autumn.

It's easy to propagate daffodils (Narcissus) from bulblets or from bulb scales.

PERENNIALS FOR FLOWERS AND FOLIAGE – *HOSTA* TO *LATHYRUS*

Plant name	Plant type	Propagation method
Hosta	Border perennials	Divide (see p98) in late summer or early spring
Houttuynia	Border perennials	Take softwood cuttings (see p44) in late spring; or divide rhizomes (see p94) in spring
Hyacinthoides (bluebell)	Perennials grown from bulbs	Divide clumps (see p98) in summer; separate bulblets (see p86) in summer; or sow seed in containers (see p126) in a cold frame as soon as ripe
Hyacinthus (hyacinth)	Perennials grown from bulbs	Separate bulblets (see p86) in summer; use bulb scoring (see p89) or scooping (see p88)
Iberis sempervirens (perennial candytuft)	Dwarf evergreen perennial	Sow seed in containers (see p126) in a cold frame in autumn; or take semiripe cuttings (see p48) in summer
Incarvillea	Border perennials	Take basal softwood cuttings (see p44) in spring; or divide (see p98) in spring
Inula (elecampane)	Border perennials	Divide (see p98) in spring or autumn
Ipheion	Dwarf perennials grown from bulbs	Divide (see p98) in summer; or sow seed in containers (see p126) in a cold frame as soon as ripe or in spring
Iris, bulbous species	Perennials grown from bulbs	Divide clumps (see p98); or separate bulblets (see p86) once leaves have died down
Iris, rhizomatous species	Border perennials	Divide rhizomes (see p94) after flowering or in early autumn
Ixia (corn lily)	Half-hardy perennials grown from corms	Separate cormels (see p91) in late summer; or sow seed in containers (see p126) in a cold frame as soon as ripe
Kirengeshoma palmata	Border perennial	Divide (see p98) as growth begins in spring
Knautia	Border perennials	Sow seed in containers (see p126) in a cold frame in spring; or take basal softwood cuttings (see p44) in spring
Kniphofia (red-hot poker)	Border perennials	Divide (see p98) in late spring; or take basal softwood cuttings (see p44) in spring. For evergreen varieties cut off the top of the crown to stimulate new shoots to use as cuttings
Lamium (dead nettle)	Border perennials	Divide (see p98) in autumn or early spring
Lamprocapnos spectabilis	Border perennial	Take root cuttings (see p74) in winter
Lathyrus, perennial species (everlasting pea)	Herbaceous perennial climbers	Soak seed for 24 hours, then sow in containers (see p126) in a cold frame in early spring

PERENNIALS FOR FLOWERS AND FOLIAGE – *LEONOTIS* TO *MACLEAYA*

Plant name	Plant type	Propagation method
Leonotis	Very tall border perennials	Sow seed (see p126) at 13–18°C (55–64°F); or take greenwood cuttings (see p47) in late spring or summer
Leucanthemum × *superbum* (Shasta daisy)	Border perennial	Divide (see p98) in early spring or late summer
Leucojum (summer snowflake)	Perennials grown from bulbs	Divide clumps (see p98) or separate bulblets (see p86) once the leaves have died down
Liatris (blazing star)	Border perennials	Divide (see p98) in spring; or sow seed in containers (see p126) in a cold frame in autumn
Ligularia	Border perennials	Divide (see p98) in spring or after flowering
Lilium (lily)	Perennials grown from bulbs	Divide clumps (see p98) or separate bulblets (see p86) after foliage dies down. Use scaling (see p87) on dormant bulbs. Detach and plant stem bulbils (see p86) from those species that produce them
Linaria, perennial species (toadflax)	Border perennials	Sow seed in containers (see p126) in a cold frame in early spring; or divide (see p98) in spring
Liriope muscari (lily turf)	Dwarf perennial	Divide (see p98) in spring
Lithodora	Dwarf perennials	Take semiripe cuttings (see p48) in summer
Lobelia, perennial species	Border perennials	Divide (see p98) in spring; or sow seed (see p126) at 13–18°C (55–64°F) as soon as ripe. Take leaf-bud cuttings (see p58) of L. cardinalis in midsummer
Lupinus (lupin)	Border perennials	Nick or soak seed for 24 hours (see p123), then sow in a seedbed (see p116) in spring or autumn; or take basal softwood cuttings (see p44) in midspring
Lychnis (campion)	Border perennials	Divide (see p98) in early spring; take basal softwood cuttings (see p44) in spring; or sow seed in containers (see p126) in a cold frame as soon as ripe or in spring
Lysimachia	Border perennials	Divide (see p98) in spring or autumn; or sow seed in containers (see p126) in a cold frame in spring
Lythrum (purple loosestrife)	Border perennials	Divide (see p98) in spring; take basal softwood cuttings (see p44) in spring or early summer; or sow seed (see p126) at 13–18°C (55–64°F) in spring
Macleaya (plume poppy)	Border perennials	Divide (see p98) by separating rooted sections from edge of clump without lifting in autumn or early spring; or take root cuttings (see p74) in winter

PERENNIALS FOR FLOWERS AND FOLIAGE – *MALVA* TO *MUSCARI*

Plant name	Plant type	Propagation method
Malva (mallow)	Border perennials	*Take basal softwood cuttings (see p44) in spring; or sow seed in situ (see p116) or in containers (see p126) in a cold frame in spring or early summer*
Meconopsis (blue poppy)	Border perennials	*Sow seed in containers (see p126) in a cold frame in spring or autumn. Divide longer-lived species (see p98) in spring*
Melissa officinalis (lemon balm)	Border perennial	*Divide (see p98) in spring or autumn; or sow seed in containers (see p126) in a cold frame in spring*
Milium effusum	Semievergreen ornamental grass	*Divide (see p98) in early spring; or sow seed in situ (see p116) in spring*
Mirabilis jalapa (marvel of Peru)	Half-hardy perennial	*Divide tubers (see p92) in spring; or sow seed in containers (see p126) at 13–18°C (55–64°F) in early spring or in situ (see p116) in late spring*
Miscanthus	Ornamental grasses	*Divide (see p98) in spring and pot up until established*
Molinia	Ornamental grasses	*Divide (see p98) in spring and pot up until established. Sow seed of species in containers (see p126) in a cold frame in spring*
Monarda (bee balm, bergamot)	Border perennials	*Divide (see p98) in spring or autumn; or root basal softwood cuttings (see p44) in spring*
Morina	Evergreen border perennials	*Take root cuttings (see p74) in winter*
Muscari (grape hyacinth)	Dwarf perennials grown from bulbs	*Divide clumps (see p98); separate bulblets (see p86) in summer; or sow seed in containers (see p126) in a cold frame in autumn*

Oriental poppies can be divided in spring or grown from root cuttings.

Ferns, such as Polypodium, are easy to multiply by division in spring or early summer.

Propagate tulips by separating the young bulblets during summer.

PERENNIALS FOR FLOWERS AND FOLIAGE – *NARCISSUS* TO *PHORMIUM*

Plant name	Plant type	Propagation method
Narcissus (daffodil)	Perennials grown from bulbs	Separate bulblets (see p86) in early summer or early autumn; or use bulb scaling (see p87) in early autumn
Nepeta (catmint)	Border perennials	Divide (see p98) in spring or autumn; or take softwood cuttings (see p44) in early summer
Nerine	Perennials grown from bulbs	Separate bulblets (see p86) after flowering
Oenothera, perennial species (evening primrose)	Border perennials	Divide (see p98) in early spring; take softwood cuttings from late spring to midsummer (see p44); or sow seed in containers (see p126) in a cold frame in early spring
Ophiopogon planiscapus (lilyturf)	Dwarf border perennial	Divide (see p98) in spring
Ornithogalum (star-of-Bethlehem)	Perennials grown from bulbs	Remove bulblets (see p86) when dormant
Osteospermum	Half-hardy and tender perennials	Sow seed in containers (see p126) at 18°C (64°F) in spring; take softwood cuttings (see p44) in late spring; or take semiripe cuttings (see p48) in late summer
Oxalis	Dwarf perennials grown from tubers	Divide (see p98) in spring
Paeonia, herbaceous species (peony)	Border perennials	Divide (see p98) in autumn or early spring
Papaver orientale (oriental poppy)	Border perennial	Divide (see p98) in spring; or take root cuttings (see p74) in late autumn or early winter
Pennisetum	Ornamental grasses	Sow seed (see p126) at 13–18°C (55–64°F) in early spring; or divide (see p98) in late spring or early summer
Penstemon	Border perennials	Take semiripe cuttings (see p48) in midsummer; take softwood cuttings (see p44) in spring; or sow seed in containers (see p126) at 13–18°C (55–64°F) in spring
Persicaria	Border perennials	Divide (see p98) in spring or autumn
Phalaris arundinacea (reed canary grass)	Ornamental grass	Divide (see p98) from midspring to midsummer
Phlox douglasii	Dwarf evergreen perennial	Take softwood cuttings (see p44) in spring
Phlox paniculata	Border perennial	Divide (see p98) in autumn or spring; or take root cuttings (see p74) in early autumn or winter
Phormium	Evergreen border perennials	Divide (see p98) in spring; or sow seed in containers (see p126) at 13–18°C (55–64°F) in spring

PERENNIALS FOR FLOWERS AND FOLIAGE – *PHYGELIUS* TO *SALVIA*

Plant name	Plant type	Propagation method
Phygelius (Cape figwort)	Half-hardy shrubs grown as border perennials	Remove rooted suckers (see p75) in spring; or take softwood cuttings (see p44) in late spring
Physalis (Chinese lantern)	Border perennials	Divide (see p98) in spring; or sow seed in containers (see p126) in a cold frame in spring
Platycodon grandiflorus (balloon flower)	Border perennial	Detach rooted basal shoots in early summer; divide (see p98) in summer; or sow seed in situ (see p116) or in containers (see p126) in a cold frame in spring
Polemonium (Jacob's ladder)	Border perennials	Divide (see p98) in spring; or sow seed in containers (see p126) in a cold frame in autumn or spring
Polygonatum (Solomon's seal)	Border perennials	Divide (see p98) when growth begins in spring
Polypodium	Evergreen ferns	Divide (see p98) in spring or early summer
Potentilla, herbaceous species (cinquefoil)	Border perennials	Divide (see p98) in autumn or spring; or take basal softwood cuttings (see p44) in midspring
Primula (primroses and polyanthus)	Border perennials	Divide (see p98) from autumn to early spring; take softwood basal cuttings (see p44) or offsets (see p100) in autumn; or sow seed in containers (see p126) in a cold frame as soon as ripe or in late winter or early spring
Prunella (self-heal)	Border perennials	Divide (see p98) in spring or autumn
Pulmonaria (lungwort)	Border perennials	Divide (see p98) after flowering or in autumn
Pulsatilla (pasque flower)	Dwarf perennials	Sow seed in containers (see p126) in a cold frame as soon as ripe; or take root cuttings (see p74) in winter
Ratibida (Mexican hat)	Border perennials	Sow seed in containers (see p126) in a cold frame in early spring; or divide young plants (see p98) in spring
Rheum palmatum (ornamental rhubarb)	Border perennial	Divide (see p98) in early spring
Rodgersia	Border perennials	Divide (see p98) in early spring
Rudbeckia, perennial species	Border perennials	Sow seed in containers (see p126) in a cold frame in early spring; or divide (see p98) in autumn or spring
Salvia, perennial species	Tender, half-hardy and hardy border perennials	Sow seed in containers (see p126) in a cold frame in spring; divide (see p98) in spring; take basal softwood cuttings (see p44) in spring; or take semiripe cuttings (see p48) in late summer or autumn with bottom heat

PERENNIALS FOR FLOWERS AND FOLIAGE – *SCABIOSA* TO *TIARELLA*

Plant name	Plant type	Propagation method
Scabiosa	Border perennials	Divide (see p98) in spring; or take basal softwood cuttings (see p44) in spring
Schizostylis coccinea	Evergreen border perennial	Divide (see p98) in spring
Scilla	Dwarf perennials grown from bulbs	Separate bulblets (see p86) in summer; or sow seed in containers (see p126) in a cold frame as soon as ripe
Sedum	Succulent border perennials, mostly evergreen	Take softwood cuttings (see p44) in early summer; or divide (see p98) in spring
Sempervivum (houseleek)	Dwarf evergreen succulents	Root offsets (see p100) in spring or early summer
Sidalcea	Border perennials	Divide (see p98) in autumn or spring
Silene (campion, catchfly)	Border perennials, many dwarf	Sow seed in containers (see p126) in a cold frame in autumn; or take basal softwood cuttings (see p44) in spring
Sisyrinchium striatum	Border perennial	Divide (see p98) in spring
Solidago (golden rod)	Border perennials	Divide (see p98) in autumn or spring
Stachys	Border perennials	Divide (see p98) in spring; or remove rooted sections of shoot in spring
Stipa	Ornamental grasses	Divide (see p98) from midspring to early summer; or sow seed in containers (see p126) in a cold frame in spring
Stokesia laevis	Border perennial	Divide (see p98) in spring; take root cuttings (see p74) in late winter; or sow seed in containers (see p126) in a cold frame in autumn
Symphytum (comfrey)	Border perennials	Divide (see p98) in spring; take root cuttings (see p74) in early winter; or sow seed in containers (see p126) in a cold frame in spring or autumn
Tanacetum	Border perennials	Sow seed (see p126) at 13–18°C (55–64°F) in late winter or early spring; divide (see p98) in spring; or take basal softwood cuttings (see p44) in spring
Tellima grandiflora	Border perennial	Divide (see p98) in spring; or sow seed in containers (see p126) in a cold frame as soon as ripe or in spring
Thalictrum (meadow rue)	Border perennials	Sow seed in containers (see p126) in a cold frame as soon as ripe or in early spring; or divide (see p98) in spring
Tiarella (foam flower)	Border perennials	Sow seed in containers (see p126) in a cold frame as soon as ripe or in spring; or divide (see p98) in spring

PERENNIALS FOR FLOWERS AND FOLIAGE – *TOLMIEA* TO *ZAUSCHNERIA*

Plant name	Plant type	Propagation method
Tolmiea menziesii (pick-a-back plant)	Border perennial	Divide (see p98) in spring; or peg leaves onto compost (see p60) and remove plantlets when rooted
Tradescantia, hardy species	Border perennials	Divide (see p98) in autumn or spring
Tricyrtis (toad lily)	Border perennials	Divide (see p98) in early spring when dormant
Trillium (wake robin)	Perennials grown from tubers	Divide rhizomes (see p94) after flowering; or cut out the growing point from the rhizome after flowering, which stimulates the formation of offsets, which can then be separated (see p100)
Triteleia	Perennials grown from corms	Separate corms (see p90) when dormant
Trollius (globeflower)	Border perennials	Divide (see p98) in spring or immediately after flowering
Tulipa (tulip)	Perennials grown from bulbs	Separate bulblets (see p86) in summer
Veratrum	Border perennials	Divide (see p98) in autumn or early spring; or sow seed in containers (see p126) in a cold frame as soon as ripe
Verbascum, perennial species (mullein)	Border perennials	Divide (see p98) in spring; take root cuttings (see p74) in autumn; or sow seed in containers (see p126) in a cold frame in late spring or early summer
Verbena, perennial species	Border perennials	Divide (see p98) in spring; take softwood cuttings (see p44) in late summer; or sow seed (see p126) at 18–21°C (64–70°F) in autumn or early spring
Veronica	Border perennials	Divide (see p98) in autumn or spring
Veronicastrum virginicum	Border perennial	Divide (see p98) in spring
Viola (violet)	Dwarf perennials	Divide (see p98) in spring or autumn; or sow seed in containers (see p126) in a cold frame as soon as ripe or in spring. (For pansies see p152.)
Waldsteinia ternata	Dwarf perennial	Divide (see p98) in early spring
Zantedeschia (arum lily)	Tender perennials grown from tubers	Divide (see p98) in spring
Zauschneria californica (Californinan fuchsia)	Border perennial	Take basal softwood cuttings in spring (see p44); or sow seed in containers in a cold frame in spring (see p126)

WOODY PLANTS FOR FLOWERS AND FOLIAGE – *ABELIA* TO *ALOYSIA*

Plant name	Plant type	Propagation method
Abelia	Shrubs grown for flowers and semievergreen foliage	*Take semiripe cuttings (see p48) in late summer or greenwood cuttings (see p47) in early summer*
Abutilon (flowering maple)	Half-hardy shrubs grown for flowers and foliage	*Take softwood cuttings (see p44) in spring or greenwood cuttings (see p47) in summer; or sow seed in containers (see p126) at 15–18°C (59–64°F) in spring*
Acacia (wattle, mimosa)	Half-hardy trees grown for fragrant flowers and foliage	*Sow seed (see p126) in spring at 18°C (64°F) or more after soaking in warm water until swollen; or take semiripe cuttings (see p48) in summer*
Acer (maple)	Trees grown for foliage	*Sow seed in situ (see p116) or in pots (see p126) outdoors as soon as ripe. Air layer Japanese maples (see p107) in spring; side-wedge graft (see p142) in late winter; or chip-bud (see p144) in late summer*
Actinida deliciosa (kiwi fruit, Chinese gooseberry)	Woody climber grown for edible fruit	*Layer (see p104) in autumn; take greenwood cuttings (see p47) in early summer; or whip-and-tongue graft (see p138) in late winter*
Actinidia kolomikta	Climber grown for foliage	*Take semiripe cuttings (see p48) in late summer*
Aesculus (buckeye, horse chestnut)	Trees and large shrubs grown for flowers	*Sow conkers in a seedbed (see p116) as soon as ripe. Separate suckers of A. parviflora (see p75). Chip-bud cultivars (see p144) from mid- to late summer*
Akebia (chocolate vine)	Climbers grown for fragrant flowers and foliage	*Layer (see p104) in winter; or take semiripe cuttings (see p48) in summer*
Aloysia citriodora (lemon verbena)	Half-hardy shrub grown for aromatic foliage	*Take softwood (see p44) or greenwood (see p47) cuttings in summer*

Named varieties of Japanese maples (Acer japonicum) are best propagated by grafting.

Sow seeds of hawthorn (Crataegus) or take chip-buds of named varieties in summer.

Ivies (Hedera) are easy to propagate by simple layering; or take semiripe cuttings.

WOODY PLANTS FOR FLOWERS AND FOLIAGE – *AMELANCHIER* TO *CALLISTEMON*

Plant name	Plant type	Propagation method
Amelanchier (snowy mespilus)	Small trees or large shrubs grown for flowers and autumn colour	Layer (see p104) in autumn; separate rooted suckers (see p75) in autumn; sow seed in a seedbed (see p116) as soon as ripe; or take semiripe cuttings (see p48) in summer
Ampelopsis	Climbers grown for foliage	Take softwood cuttings (see p44) in summer; or sow seed in pots (see p126) outdoors in autumn
Aralia (angelica tree)	Shrubs grown for foliage	Transplant rooted suckers (see p75) in spring; or take root cuttings (see p74) in winter
Arbutus (strawberry tree)	Trees grown for flowers, fruit, evergreen foliage and bark	Sow seed in pots (see p126) in a cold frame as soon as ripe; or take semiripe cuttings (see p48) in late summer
Aristolochia (Dutchman's pipe)	Climbers grown for flowers and foliage	Sow seed (see p126) of hardy species at 13–16°C (55–61°F) and tender species at 21–24°C (70–75°F) as soon as ripe or in spring; or take softwood cuttings (see p44) of hardy species in midsummer, tender species in early spring
Artemisia	Shrubs grown for aromatic evergreen foliage	Take greenwood cuttings (see p47) or heel cuttings (see p56) of sideshoots in early summer
Aucuba (spotted laurel)	Shrubs grown for evergreen foliage	Take semiripe cuttings (see p48) in summer
Ballota	Subshrubs grown for evergreen foliage	Take semiripe cuttings (see p48) in early summer or softwood cuttings (see p44) in late spring
Berberis	Shrubs grown for flowers, foliage and berries	Take semiripe (see p48) heel or mallet (see p56) cuttings of deciduous types in summer. Take evergreen cuttings (see p54) of evergreens in autumn
Betula (birch)	Trees grown for bark, foliage and habit	Sow seed of species in a seedbed (see p116) in autumn. Side-wedge graft cultivars (see p142) in winter
Brachyglottis	Shrubs grown for flowers and evergreen foliage	Take semiripe cuttings (see p48) in summer
Brugmansia (angels' trumpets)	Tender shrubs grown for flowers	Take semiripe cuttings (see p48) in summer and root with bottom heat; or sow seed in containers (see p126) at 16°C (61°F) in spring
Buddleja (butterfly bush)	Hardy and half-hardy shrubs grown for flowers	Take semiripe cuttings (see p48) in summer. Take hardwood cuttings (see p50) of B. davidii in autumn
Buxus (box)	Shrubs grown for evergreen foliage	Take semiripe cuttings (see p48) in summer
Callistemon, hardy species (bottlebrush)	Shrubs grown for flowers	Sow seed (see p126) on the surface of moist compost at 16–18°C (61–64°F) in spring; or take semiripe cuttings (see p48) in late summer and root with bottom heat

WOODY PLANTS FOR FLOWERS AND FOLIAGE – *CALLUNA* TO *COPROSMA*

Plant name	Plant type	Propagation method
Calluna vulgaris (heather)	Shrub grown for flowers and evergreen foliage	Take semiripe cuttings (see p48) in midsummer; layer (see p104) in spring; or layer by dropping (see p112) in spring
Camellia	Shrubs grown for flowers and evergreen foliage	Layer (see p104) in late winter or early spring; take leaf-bud (see p58) or semiripe (see p48) cuttings of the current year's growth from late summer to late winter. Apical-wedge graft (see p140) *C. reticulata* cultivars from mid- to late winter
Campsis (trumpet vine)	Climbers grown for flowers	Layer (see p104) in autumn; take leaf-bud cuttings (see p58) in spring; semiripe cuttings (see p48) in summer; or hardwood cuttings (see p50) in winter
Carpinus (hornbeam)	Trees grown for foliage, habit and as hedging	Sow seeds of species in a seedbed (see p116) in autumn. Take greenwood cuttings (see p47) of cultivars in early summer; or whip-and-tongue graft (see p138) in winter
Caryopteris × *clandonensis*	Shrub grown for flowers	Take softwood cuttings (see p44) in late spring or greenwood cuttings (see p47) in early summer
Catalpa (Indian bean tree)	Trees grown for foliage, flowers and seed pods	Sow seed in a seedbed (see p116) in autumn; take softwood cuttings (see p44) in late spring or summer; or chip-bud (see p144) in summer
Ceanothus	Shrubs grown for flowers	Take greenwood heel cuttings (see p47) of deciduous types from mid- to late summer. Root semiripe heel cuttings of evergreen varieties (see p54) from mid- to late summer
Ceratostigma willmottianum	Shrub grown for flowers	Layer (see p104) in autumn; or take semiripe cuttings (see p48) in summer
Chaenomeles (flowering quince, japonica)	Shrubs grown for flowers and fruit	Layer (see p104) in autumn; or take semiripe cuttings (see p48) in summer
Chamaecyparis (cypress)	Conifers grown for evergreen foliage, habit and as hedging	Take semiripe cuttings (see p48) in late summer. Side-wedge graft (see p142) dwarf cultivars of *C. obtusa* in winter
Choisya (Mexican orange blossom)	Shrubs grown for flowers and evergreen foliage	Take semiripe cuttings (see p48) in summer
Cistus (sun rose)	Shrubs grown for flowers and evergreen foliage	Take softwood (see p44) or greenwood (see p47) cuttings in summer
Clematis	Climbers grown for flowers	Sow seed of species as soon as ripe in pots (see p126) in a cold frame. Layer (see p104) species and cultivars in late winter or early spring; or take leaf-bud cuttings (see p58) from early to midsummer
Coprosma	Shrubs grown for evergreen foliage	Take semiripe cuttings (see p48) in late summer

WOODY PLANTS FOR FLOWERS AND FOLIAGE – *CORDYLINE* TO *CRATAEGUS*

Plant name	Plant type	Propagation method
Cordyline (New Zealand flax)	Half-hardy to tender shrubs grown for evergreen foliage	Separate well-rooted suckers (see p75) in spring; or sow seed in containers (see p126) at 16°F (61°C) in spring
Cornus, shrub species (dogwood)	Shrubs mostly grown for foliage and winter-stem colour	Layer (see p104) in spring; or take hardwood cuttings (see p50) in winter
Cornus, tree species (flowering dogwood)	Trees grown for flower-like bracts	Sow seed of species in a seedbed (see p116) in autumn; or stratify (see p125) and sow in spring. Side-wedge graft (see p142) cultivars in winter
Corylus (hazel)	Shrubs and trees grown for catkins, foliage and nuts	Sow nuts of species in a seedbed (see p116) as soon as ripe. Apical-wedge graft (see p140) C. avellana 'Contorta' and C. a. 'Pendula' in winter. Layer (see p104) other cultivars in autumn; or side-wedge graft (see p142) in winter
Cotinus (smoke bush)	Shrubs grown for flowers and foliage	Layer (see p104) in spring
Cotoneaster	Shrubs grown for flowers, berries and foliage	Sow berries of species in pots (see p126) in a cold frame as soon as ripe. Take greenwood cuttings (see p47) of deciduous types in early summer. Take semiripe cuttings (see p48) of evergreens and semievergreens in late summer
Crataegus (hawthorn)	Trees grown for flowers, berries, autumn colour and as hedging	Remove flesh and sow seed of species as soon as ripe in a seedbed (see p116) or in pots (see p126) outdoors; or stratify (see p125) in winter and sow in spring. Chip-bud (see p144) cultivars in midsummer

Propagate Magnolia grandiflora *by layering, semiripe cuttings or grafting.*

Passion flowers (Passiflora) *are easy to propagate by layering or from semiripe cuttings.*

Named cultivars of cherries (Prunus) *should be propagated by grafting.*

WOODY PLANTS FOR FLOWERS AND FOLIAGE – *CUPRESSUS* TO *EXOCHORDA*

Plant name	Plant type	Propagation method
Cupressus (cypress)	Conifers grown for evergreen foliage, habit and as hedging	*Take semiripe cuttings (see p48) in late summer*
× *Cuprocyparis* (cypress)	Conifers grown for evergreen foliage, habit and as hedging	*Take semiripe cuttings (see p48) in late summer*
Cytisus (broom)	Shrubs grown for flowers	*Sow seed in pots (see p126) in a cold frame in autumn or spring; or take semiripe cuttings (see p48) in late summer*
Daphne	Shrubs grown for fragrant flowers and evergreen foliage	*Layer (see p104) in spring; take semiripe cuttings (see p48) in mid- or late summer; sow seed in pots (see p126) in a cold frame as soon as ripe; or apical-wedge (see p140), side-wedge (see p142) or whip-and-tongue (see p138) graft in winter*
Davidia involucrata (dove tree, handkerchief tree)	Tree grown for flower-like bracts	*Sow the whole fruit in pots (see p126) outdoors as soon as ripe – it requires two winters to germinate*
Deutzia	Shrubs grown for flowers	*Take hardwood cuttings (see p50) in autumn or softwood cuttings (see p44) in summer*
Elaeagnus	Shrubs grown for foliage, many evergreen	*Take semiripe cuttings (see p48) in midsummer; or separate rooted suckers (see p75) of deciduous species in autumn*
Enkianthus	Shrubs grown for flowers and evergreen foliage	*Layer (see p104) in autumn; or take semiripe cuttings (see p48) in spring*
Erica (heath)	Shrubs grown for flowers and evergreen foliage	*Take semiripe cuttings (see p48) in midsummer; layer (see p104) in spring; or layer by dropping (see p112) in spring*
Escallonia	Shrubs grown for flowers and evergreen foliage	*Take evergreen cuttings (see p54) in late autumn or winter or semiripe cuttings (see p48) in late summer*
Eucalyptus (gum tree)	Trees grown for aromatic foliage and bark	*Sow seed at 13–18°C (55–64°F) in containers (see p126) in spring or summer*
Eucryphia	Trees grown for flowers and evergreen foliage	*Take semiripe cuttings (see p48) in summer. Sow seed of species in containers (see p126) in a cold frame as soon as ripe or in late winter, then overwinter young plants under glass*
Euonymus (spindle)	Shrubs grown for evergreen foliage or berries and autumn colour	*Take semiripe cuttings (see p48) of evergreens in summer. Take greenwood cuttings (see p47) of deciduous types in summer. Sow seed of* E. europaeus *in pots (see p126) in a cold frame as soon as ripe*
Exochorda (pearl bush)	Shrubs grown for flowers	*Take softwood cuttings (see p44) in summer and root with bottom heat*

WOODY PLANTS FOR FLOWERS AND FOLIAGE – *FAGUS* TO *HAMAMELIS*

Plant name	Plant type	Propagation method
Fagus (beech)	Trees grown for habit and foliage	Sow seed in a seedbed (see p116) in autumn; or stratify (see p125) in winter and sow in spring. Whip-and-tongue (see p138) or side-wedge (see p142) graft cultivars in midwinter
Fargesia	Evergreen bamboo	Divide established clumps (see p98) in spring; or take cuttings from young rhizomes (see p94) in spring
Fatsia japonica (Japanese aralia)	Half-hardy shrub grown for evergreen foliage, flowers and berries	Air layer (see p107) in spring or late summer; or take greenwood cuttings (see p47) in early or midsummer
Ficus carica (fig)	Tree or large shrub grown for edible fruit	Root semiripe cuttings (see p48) or leaf-bud cuttings (see p58) with bottom heat in spring or summer
Forsythia	Shrubs grown for flowers	Take semiripe cuttings (see p48) in late summer or hardwood cuttings (see p50) in midautumn
Fothergilla	Shrubs grown for flowers and autumn colour	Air layer (see p107) in spring; or layer (see p104) in summer
Fraxinus (ash)	Trees grown for foliage and habit	Sow seed in pots (see p126) in a cold frame as soon as ripe. Whip-and-tongue graft (see p138) cultivars in late winter or early spring
Fremontodendron	Shrubs grown for flowers and evergreen foliage	Take semiripe cuttings (see p48) in late summer or greenwood cuttings (see p47) in early summer
Fuchsia	Hardy to tender shrubs grown for flowers	Take softwood cuttings (see p44) in spring or semiripe cuttings (see p48) in summer
Garrya (silk tassel bush)	Shrubs grown for catkins and evergreen foliage	Take evergreen cuttings (see p54) in autumn or semiripe cuttings (see p48) in summer
Gaultheria	Shrubs grown for flowers, berries and evergreen foliage	Separate rooted suckers (see p75) in spring; take semiripe cuttings (see p48) in summer; or sow seed in pots (see p126) in a cold frame in autumn
Genista (broom)	Shrubs grown for flowers	Sow seed in pots (see p126) in a cold frame in autumn or spring; or take semiripe cuttings (see p48) in summer
Ginkgo biloba (maidenhair tree)	Tree grown for foliage and habit	Sow seed in pots (see p126) outdoors as soon as ripe; or whip-and-tongue graft (see p138) in winter
Gleditsia	Trees grown for foliage, flowers and seed pods	Scarify seed (see p124) and sow in pots (see p126) outdoors in autumn. Whip-and-tongue graft (see p138) cultivars in late winter
Griselinia	Shrubs grown for evergreen foliage	Take semiripe cuttings (see p48) in summer
Hamamelis (witch hazel)	Shrubs grown for fragrant flowers and autumn colour	Layer (see p104) in spring; or side-wedge graft (see p142) or chip-bud (see p144) cultivars in late summer

WOODY PLANTS FOR FLOWERS AND FOLIAGE – *HEBE* TO *LABURNUM*

Plant name	Plant type	Propagation method
Hebe	Shrubs grown for flowers and evergreen foliage	Take semiripe cuttings (see p48) in late summer or autumn or softwood cuttings (see p44) in late spring
Hedera (ivy)	Climbers grown for evergreen foliage	Layer (see p104) in spring or autumn; or take semiripe cuttings (see p48) in summer
Helianthemum (rock rose)	Shrubs grown for flowers	Take softwood cuttings (see p44) in late spring or early summer
Hibiscus syriacus	Shrub grown for flowers	Take semiripe cuttings (see p48) in summer or greenwood cuttings (see p47) in late spring
Hydrangea	Shrubs and climbers grown for flowers and, in some species, foliage	Take softwood cuttings (see p44) of deciduous varieties in spring or hardwood cuttings (see p50) in winter. Take semiripe cuttings of evergreens (see p54) in summer and root with bottom heat
Hypericum (St John's wort)	Shrubs grown for flowers	Take greenwood (see p47) or semiripe (see p48) cuttings in summer
Ilex (holly)	Shrubs grown for evergreen foliage and berries	Take semiripe cuttings (see p48) in late summer or early autumn; or sow seed in pots (see p126) in a cold frame in autumn
Itea	Shrubs grown for flowers and evergreen foliage	Take greenwood cuttings (see p47) in spring; or take semiripe cuttings (see p48) in summer and root with bottom heat
Jasminum, hardy species (jasmine)	Shrubs and climbers grown for fragrant flowers and mostly evergreen leaves	Layer (see p104) in autumn; or take semiripe cuttings (see p48) in summer or hardwood cuttings (see p50) or evergreen cuttings (see p54) in late autumn
Juglans (walnut)	Trees grown for nuts, foliage and habit	Sow nuts in a seedbed (see p116) as soon as ripe; or stratify (see p125) over winter and sow in spring
Juniperus (juniper)	Conifers grown for foliage and habit	Take hardwood cuttings (see p50) in early autumn
Kerria japonica	Shrub grown for flowers and foliage	Divide (see p98) in autumn; or take greenwood cuttings (see p47) in summer
Koelreuteria	Trees grown for flowers, foliage and seed pods	Sow seed in pots (see p126) in a cold frame in autumn; or take root cuttings (see p74) in winter
Kolkwitzia amabilis (beauty bush)	Shrub grown for flowers	Separate suckers (see p75) in spring; or take greenwood cuttings (see p47) in late spring or early summer
Laburnum	Trees grown for flowers	Take hardwood cuttings (see p50) in winter. Sow seed of species in pots (see p126) in a cold frame in autumn. Apical-wedge graft (see p140) cultivars in early spring

WOODY PLANTS FOR FLOWERS AND FOLIAGE – *LAPAGERIA* TO *MAHONIA*

Plant name	Plant type	Propagation method
Lapageria rosea (Chilean bellflower)	Half-hardy climber grown for flowers	Layer (see p104) in autumn; take semiripe cuttings (see p48) in late summer; or soak seed for 48 hours, then sow in containers (see p126) at 13–18°C (55–64°F) in spring
Laurus nobilis (bay)	Tree grown for aromatic evergreen foliage	Take semiripe cuttings (see p48) in summer; or sow seeds in containers (see p126) in a cold frame in autumn
Lavandula (lavender)	Shrubs grown for fragrant flowers and aromatic evergreen foliage	Take semiripe cuttings (see p48) in summer
Lavatera, shrubby species (mallow)	Shrubs grown for flowers	Take softwood (see p44) or greenwood (see p47) cuttings in early summer
Leptospermum (tea tree)	Shrubs grown for flowers and aromatic foliage	Take semiripe cuttings (see p48) in summer and root with bottom heat; or sow seed at 13–16°C (55–61°F) in containers (see p126) in autumn or spring
Leucothoe	Shrubs grown for flowers and evergreen foliage	Take semiripe cuttings (see p48) in summer and root with bottom heat. Divide suckering species (see p75) in spring
Ligustrum (privet)	Shrubs and trees grown for evergreen or semievergreen foliage, flowers and as hedging	Take hardwood or evergreen cuttings (see p50) in winter or semiripe cuttings (see p48) in summer
Liquidambar (sweet gum)	Trees grown for foliage and habit	Sow seed in pots (see p126) in a cold frame in autumn; layer (see p104) in autumn; or take greenwood cuttings (see p47) in summer
Lonicera (honeysuckle)	Shrubs and climbers grown for fragrant flowers and, in some species, evergreen foliage	Layer (see p104) in spring. Take semiripe cuttings of evergreens (see p54) in summer. Take hardwood cuttings (see p50) of deciduous types in autumn
Lotus hirsutus (hairy Canary clover)	Subshrub grown for flowers and evergreen foliage	Take semiripe cuttings (see p48) in summer
Luma (myrtle)	Shrubs grown for fragrant flowers and aromatic evergreen foliage	Sow seed in pots (see p126) in a cold frame in spring; or take semiripe cuttings (see p48) in late summer
Magnolia	Trees and shrubs grown for flowers	Layer (see p104) in early spring. Take softwood cuttings (see p44) of deciduous species in late spring. Take semiripe cuttings of evergreens (see p54) in late summer; chip-bud (see p144) from mid- to late summer; or side-wedge graft (see p142) in late winter or early spring
Mahonia	Shrubs grown for fragrant flowers and evergreen foliage	Take semiripe cuttings (see p48) or leaf-bud cuttings (see p58) in late summer or autumn

WOODY PLANTS FOR FLOWERS AND FOLIAGE – *MALUS* TO *OLEA*

Plant name	Plant type	Propagation method
Malus, ornamental species (crab apple)	Trees grown for flowers, fruit and autumn colour	Whip-and-tongue graft (see p138) in late winter; or chip-bud (see p144) from mid- to late summer. M. baccata, M. bhutanica, M. florentina, M. hupehensis *and* M. sikkimensis *come true from seed sown outdoors (see p116) as soon as ripe*
Malus domestica (apple)	Tree grown for edible fruit	Chip-bud (see p144) from mid- to late summer; or whip-and-tongue graft (see p138) in late winter
Morus (mulberry)	Trees grown for fruit, foliage and habit	Take semiripe cuttings (see p48) in summer; or take hardwood cuttings (see p50) in autumn and root in a prepared bed in a cold frame; two-to-four-year-old branches, or truncheons, will also root if treated like hardwood cuttings. Chip-bud (see p144) cultivars in late summer
Myrtus (myrtle)	Shrubs grown for fragrant flowers and aromatic foliage	Sow seed in pots (see p126) in a cold frame in autumn; or take semiripe cuttings (see p48) in autumn and root with bottom heat
Nandina domestica (heavenly bamboo)	Shrub grown for flowers, foliage and berries	Take semiripe cuttings (see p48) in summer
Nerium (oleander)	Tender shrubs grown for flowers and evergreen foliage	Take semiripe cuttings (see p48) in summer; or air layer (see p107) in spring
Olea europaea (olive)	Half-hardy tree grown for foliage, fruit and habit	Sow seed in containers (see p126) at 13–15°C (55–59°F) in spring; or take semiripe cuttings (see p48) in summer

Grow roses (Rosa) on their own roots by taking hardwood cuttings or try chip-budding.

Propagate rosemary (Rosmarinus) from semiripe cuttings taken in summer.

Variegated sage (Salvia) can be propagated from softwood or semiripe cuttings.

WOODY PLANTS FOR FLOWERS AND FOLIAGE – *OLEARIA* TO *PICEA*

Plant name	Plant type	Propagation method
Olearia, hardy species (daisy bush)	Shrubs and small trees grown for flowers and evergreen foliage	*Take semiripe cuttings (see p48) in summer*
Osmanthus	Shrubs grown for fragrant flowers and evergreen foliage	*Layer (see p104) in autumn or spring; or take semiripe cuttings (see p48) in summer and root with bottom heat*
Pachysandra terminalis	Shrub grown for evergreen foliage	*Divide (see p98) in spring*
Paeonia, shrubby species (tree peony)	Shrubs grown for flowers and foliage	*Take semiripe cuttings (see p48) in summer. Sow seed of species in pots (see p126) outdoors in autumn or early winter, which may take two or three years to germinate. Apical-wedge graft (see p140) cultivars in winter*
Parrotia persica (Persian ironwood)	Shrubs grown for autumn colour, bark and flowers	*Layer (see p104) in spring; or take semiripe cuttings (see p48) from mid- to late summer*
Parthenocissus (Virginia creeper)	Climbers grown for foliage	*Layer (see p104) from spring to autumn; sow seed in pots (see p126) in a cold frame in autumn; or take hardwood cuttings (see p50) in winter or softwood cuttings (see p44) in early summer*
Passiflora (passion flower)	Hardy to tender climbers grown for flowers, fruit and foliage	*Layer (see p104) in spring or autumn; or take semiripe cuttings (see p48) in summer*
Paulownia	Trees grown for flowers and foliage	*Sow seed in pots (see p126) in a cold frame in autumn or spring; or take root cuttings (see p74) in winter and overwinter young plants under glass*
Perovskia	Subshrubs grown for flowers and aromatic foliage	*Take semiripe cuttings (see p48) in summer*
Philadelphus (mock orange)	Shrubs grown for fragrant flowers	*Take hardwood cuttings (see p50) in autumn or winter or softwood cuttings (see p44) in summer*
Photinia	Trees and shrubs grown for evergreen foliage, flowers and berries	*Take semiripe cuttings (see p48) and root with bottom heat in summer; or sow seed in pots (see p126) in a cold frame in autumn*
Phyllostachys	Evergreen bamboos	*Divide (see p98) in spring*
Physocarpus opulifolius (ninebark)	Shrub grown for foliage and flowers	*Separate rooted suckers (see p75) in autumn or spring; or take greenwood cuttings (see p47) in summer*
Picea (spruce)	Conifers grown for foliage, habit and cones	*Sow seed in pots (see p126) in a cold frame in spring. Take hardwood cuttings (see p50) of dwarf cultivars in late summer. Side-wedge graft (see p142) cultivars in winter*

WOODY PLANTS FOR FLOWERS AND FOLIAGE – *PIERIS* TO *PRUNUS*

Plant name	Plant type	Propagation method
Pieris	Shrubs grown for evergreen foliage and flowers	*Layer (see p104) in autumn; take greenwood cuttings (see p47) in early summer; or semiripe cuttings (see p48) from mid- to late summer and root with bottom heat*
Pinus (pine)	Conifers grown for foliage, habit and cones	*Sow seed of species in pots (see p126) in a cold frame in spring. Side-wedge graft (see p142) cultivars in late winter*
Piptanthus	Shrubs grown for flowers and foliage	*Take semiripe heel cuttings (see p48) in spring or autumn; or sow seed in pots (see p126) in a cold frame in spring or autumn*
Pittosporum	Trees and shrubs grown for evergreen foliage and, in some species, fragrant flowers	*Take semiripe cuttings (see p48) in summer; layer (see p104) in spring; air layer (see p107) in spring; or sow seed in pots (see p126) in a cold frame as soon as ripe*
Pleioblastus	Evergreen bamboos	*Divide (see p98) in spring*
Plumbago auriculata (cape leadwort)	Climber grown for flowers	*Take semiripe cuttings (see p48) in summer and root with bottom heat*
Populus (aspen, poplar)	Trees grown for foliage and habit	*Take hardwood cuttings (see p50) in winter; or separate suckers (see p75) in autumn or late winter*
Potentilla, shrubby species	Shrubs grown for flowers	*Take greenwood cuttings (see p47) in early summer*
Prostanthera (mint bush)	Shrubs grown for aromatic evergreen foliage and flowers	*Take semiripe cuttings (see p48) in summer; or sow seed in containers (see p126) at 13–18°C (55–64°F) in spring*
Prunus, deciduous ornamental species	Trees and shrubs grown for flowers, foliage and habit	*Stratify (see p125) fruit of P. avium, P. padus and P. sargentii in autumn, then separate seed and sow in pots (see p126) in a cold frame in spring. Whip-and-tongue graft (see p138) cultivars in late winter or early spring; or chip-bud (see p144) in mid- or late summer*
Prunus, evergreen species	Shrubs grown for evergreen foliage and as hedging	*Take semiripe cuttings (see p48) in midsummer and root with bottom heat*
Prunus armeniaca (apricot)	Tree grown for edible fruit	*Chip-bud (see p144) in summer; or whip-and-tongue graft (see p138) in early spring*
Prunus avium, P. cerasus and *P. × gondouinii* (cherry)	Trees grown for edible fruit	*Chip-bud (see p144) from midsummer to early autumn; or whip-and-tongue graft (see p138) from late winter to early spring*
Prunus domestica (plum, damson and gage)	Tree grown for edible fruit	*Chip-bud (see p144) from midsummer to early autumn; or whip-and-tongue graft (see p138) from late winter to early spring*

WOODY PLANTS FOR FLOWERS AND FOLIAGE – *PRUNUS* TO *RIBES*

Plant name	Plant type	Propagation method
Prunus persica (peach and nectarine)	Tree grown for edible fruit	Chip-bud (see p144) from midsummer to early autumn; or whip-and-tongue graft (see p138) from late winter to early spring
Pyracantha (firethorn)	Shrubs grown for berries, flowers and semievergreen foliage	Take semiripe cuttings (see p48) in summer and root with bottom heat
Pyrus (ornamental pear)	Trees grown for flowers, foliage and habit	Chip-bud (see p144) from mid- to late summer; or whip-and-tongue graft (see p138) in early spring
Pyrus communis (pear)	Tree grown for edible fruit	Chip-bud (see p144) from mid- to late summer; or whip-and-tongue graft (see p138) in early spring
Quercus (oak)	Trees grown for foliage and habit	Sow acorns in pots (see p126) in a cold frame as soon as ripe. Side-wedge graft (see p142) cultivars in late winter
Rhamnus (buckthorn)	Shrubs grown for foliage and berries	Take greenwood cuttings (see p47) of deciduous species; or layer (see p104) in spring. Take semiripe cuttings of evergreen species (see p54) in early summer; or sow seed in pots (see p126) in a cold frame as soon as ripe
Rhododendron	Shrubs grown for flowers and evergreen foliage	Layer (see p104) in autumn; air layer (see p107) in spring; or take semiripe cuttings (see p48) in late summer
Rhus (sumach)	Shrubs and trees grown for foliage	Separate suckers (see p75) from late autumn to early spring; or take root cuttings (see p74) in winter
Ribes (flowering currant)	Shrubs grown for flowers	Take hardwood cuttings (see p50) of deciduous types in winter. Take semiripe cuttings of evergreens (see p54) in summer

Lilacs (Syringa) can be air or simple layered, and you can chip-bud them too.

Sow seed of thyme (Thymus) in spring or take semiripe cuttings in summer.

Propagate ornamental vines (Vitis) by layering or take hardwood cuttings.

WOODY PLANTS FOR FLOWERS AND FOLIAGE – *RIBES* TO *SALVIA*

Plant name	Plant type	Propagation method
Ribes nigrum (black currant)	Shrub grown for edible fruit	*Take hardwood cuttings (see p50) from late autumn to midwinter*
Ribes rubrum (red currant and white currant)	Shrub grown for edible fruit	*Take hardwood cuttings (see p50) from late autumn to midwinter*
Ribes uva-crispa (gooseberry)	Shrub grown for edible fruit	*Take hardwood cuttings (see p50) from late autumn to midwinter*
Robinia	Trees grown for foliage and flowers	*Separate suckers (see p75) in autumn; take root cuttings (see p74) in winter; or sow seed in pots (see p126) in a cold frame in autumn. Chip-bud (see p144) cultivars in early spring*
Romneya coulteri (tree poppy)	Subshrub grown for flowers and foliage	*Take root cuttings (see p74) in winter or basal softwood cuttings (see p44) in spring*
Rosa (rose)	Shrubs and climbers grown for fragrant flowers and, in some species, hips or foliage	*Take hardwood cuttings (see p50) in autumn; or chip-bud (see p144) in midsummer*
Rosmarinus officinalis (rosemary)	Shrub grown for aromatic evergreen foliage	*Take semiripe cuttings (see p48) in summer*
Rubus (blackberry and hybrid berries)	Woody climbers grown for edible fruit	*Tip layer (see p110) in late summer; or take leaf-bud cuttings (see p58) from mid- to late summer*
Rubus (ornamental bramble)	Shrubs and climbers grown for flowers and foliage and, in some species, winter-stem colour	*Take hardwood cuttings (see p50) of deciduous types in early winter. Take semiripe cuttings of evergreens (see p54) in summer. Tip layer (see p110) R. tricolor at any time.*
Rubus idaeus (raspberry)	Shrub grown for edible fruit	*Divide (see p98) from autumn to early spring*
Ruscus (butcher's broom)	Shrubs grown for evergreen leaf-like shoots and berries	*Divide (see p98) in spring; or sow seed in pots (see p126) in a cold frame as soon as ripe*
Ruta (rue)	Subshrubs grown for aromatic foliage	*Take semiripe cuttings (see p48) in summer; or sow seed in pots (see p126) in a cold frame in spring*
Salix (willow)	Trees and shrubs grown for foliage, habit, catkins and winter-stem colour	*Take hardwood cuttings (see p50) in winter or greenwood cuttings (see p47) in early summer*
Salvia officinalis (sage)	Shrub grown for aromatic evergreen foliage	*Take semiripe cuttings (see p48) in late summer or early autumn and root with bottom heat; or take softwood cuttings (see p44) in spring or early summer*

WOODY PLANTS FOR FLOWERS AND FOLIAGE – *SAMBUCUS* TO *TAXUS*

Plant name	Plant type	Propagation method
Sambucus (elder)	Shrubs grown for flowers, foliage and berries	Take hardwood cuttings (see p50) in winter or greenwood cuttings (see p47) in early summer
Santolina	Shrubs grown for flowers and aromatic evergreen foliage	Take semiripe cuttings (see p48) in summer and root with bottom heat
Sarcococca (Christmas box, sweet box)	Shrubs grown for fragrant flowers and evergreen foliage	Separate suckers (see p75) in winter; or take semiripe cuttings (see p48) in late summer
Sasa	Broad-leaved bamboos	Divide (see p98) in spring; or cut sections from youngest rhizomes (see p94) in spring
Schisandra	Climbers grown for flowers and fruit	Take semiripe cuttings (see p48) in summer or greenwood cuttings (see p47) in early or midsummer
Schizophragma	Woody climbers grown for flowers	Take semiripe cuttings (see p48) in late summer or greenwood cuttings (see p47) in early or midsummer; or layer (see p104) in spring
Skimmia	Shrubs grown for berries, flowers and evergreen foliage	Take semiripe cuttings (see p48) in late summer and root with bottom heat
Solanum	Tender to hardy shrubs and climbers grown for flowers or berries	Sow seed of tender shrubs in containers (see p126) at 18–20°C (64–68°F). Take semiripe cuttings (see p48) of shrubs and climbers from summer to early autumn and root with bottom heat
Sorbus (mountain ash, rowan, whitebeam)	Trees grown for berries, foliage and flowers	Remove seed from berries and sow as soon as ripe in pots (see p126) in a cold frame. Chip-bud cultivars from mid- to late summer (see p144)
Spiraea	Shrubs grown for flowers and foliage	Take greenwood cuttings (see p47) in summer. Divide suckering species (see p75) in late autumn or early spring
Symphoricarpos (snowberry)	Shrubs grown for berries	Divide (see p98) in autumn; or take hardwood cuttings (see p50) in autumn
Syringa (lilac)	Trees and shrubs grown for fragrant flowers	Layer (see p104) or air layer (see p107) in early summer; apical-wedge graft (see p140) in late winter; or chip-bud (see p144) from mid- to late summer
Tamarix (tamarisk)	Shrubs grown for flowers and foliage	Take hardwood cuttings (see p50) in winter or semiripe cuttings (see p48) in summer
Taxus (yew)	Conifers grown for habit, evergreen foliage, fruit and as hedging	Take semiripe cuttings (see p48) from upright shoots in late summer or early autumn. Sow seed of species in pots (see p126) in a cold frame as soon as ripe. Side-wedge graft (see p142) cultivars in late summer or late winter

WOODY PLANTS FOR FLOWERS AND FOLIAGE – *TEUCRIUM* TO *YUCCA*

Plant name	Plant type	Propagation method
Teucrium	Shrubs grown for aromatic foliage and flowers	*Take semiripe cuttings (see p48) in midsummer and root with bottom heat; or sow seed in pots (see p126) in a cold frame as soon as ripe*
Thuja (arborvitae)	Conifers grown for evergreen foliage, habit and as hedging	*Take semiripe cuttings (see p48) in late summer; or sow seed in spring in pots (see p126) in a cold frame*
Thymus (thyme)	Shrubs grown for aromatic evergreen foliage	*Divide (see p98) in spring; or take semiripe cuttings (see p48) from mid- to late summer. Sow seed of species in pots (see p126) in a cold frame in spring*
Trachelospermum	Climbers grown for fragrant flowers and evergreen foliage	*Layer (see p104) in autumn; or take semiripe cuttings (see p48) in summer and root with bottom heat*
Trachycarpus fortunei (Chusan palm)	Half-hardy palm	*Sow seed in containers (see p126) at 24°C (75°F) in spring or autumn*
Vaccinium (bilberry)	Shrubs grown for berries, foliage and flowers	*Layer (see p104) in late summer; or sow seed in pots (see p126) in a cold frame in autumn*
Vaccinium corymbosum (blueberry)	Shrub grown for edible fruit	*Take softwood cuttings (see p44) in late spring or greenwood cuttings (see p47) in midsummer and root with bottom heat*
Viburnum	Shrubs grown for flowers (often fragrant), berries and foliage	*Take greenwood cuttings (see p47) of deciduous types. Root semiripe cuttings of evergreens (see p54) in summer*
Vinca (periwinkle)	Shrubs and subshrubs with trailing stems grown for evergreen foliage and flowers	*Divide (see p98) from autumn to spring; or take semiripe cuttings (see p48) in summer*
Vitis (ornamental vines)	Woody climbers grown for foliage and fruit	*Take hardwood cuttings in late winter (see p50); or layer in autumn (see p104)*
Weigela	Shrubs grown for flowers and foliage	*Take hardwood cuttings in autumn or winter (see p50); or semiripe cuttings in midsummer (see p48)*
Wisteria	Woody climbers grown for fragrant flowers	*Layer in summer (see p104); or take basal softwood cuttings (see p44) from early to midsummer and root with bottom heat*
Yucca	Shrubs grown for evergreen foliage and flowers	*Separate offsets in spring (see p100); or take root cuttings in winter (see p74)*

INDOOR PLANTS FOR FLOWERS AND FOLIAGE – *ABUTILON* TO *BEGONIA*

Plant name	Plant type	Propagation method
Abutilon (flowering maple)	Tender shrubs grown for flowers and foliage	Root softwood cuttings (see p44) in spring or greenwood cuttings (see p47) in summer; or sow seed in containers (see p126) at 15–18°C (59–64°F) in spring
Achimenes (hot-water plant)	Tender perennials grown for flowers	Divide (see p98) in spring; or take stem cuttings (see p44) in spring
Adiantum (maidenhair fern)	Tender ferns	Divide (see p98) in early spring
Aechmea (urn plant)	Bromeliads grown for flowers and foliage	Separate offsets (see p100) in early summer
Aglaonema (Chinese evergreen)	Tender perennials grown for foliage	Divide (see p98) in spring
Aloe	Rosette-forming succulents grown for foliage and flowers	Separate offsets (see p100) in late spring or early summer
Ananas (pineapple)	Bromeliads grown for flowers, foliage and fruit	Root offsets (see p100) in early summer; or slice off the leafy rosette at the top of the fruit, allow cut surface to dry, then root with bottom heat
Anthurium (flamingo flower)	Tender perennials grown for flowers and foliage	Divide rootstock (see p98) in winter
Aphelandra squarrosa (zebra plant)	Tender perennial grown for flowers and foliage	After flowering, cut back the main stem to a pair of strong leaves to encourage sideshoots; use these as softwood cuttings (see p44) and root with bottom heat
Asparagus (asparagus fern)	Tender fern-like perennials	Sow seed in containers (see p126) at 16°C (61°F) in autumn or early spring; or divide clusters of tubers (see p82) in early spring
Aspidistra (cast-iron plant)	Half-hardy perennials grown for foliage	Divide (see p98) in spring
Asplenium	Hardy and tender ferns	Divide hardy species (see p98) in spring. A. bulbiferum produces plantlets on the leaves, which can be detached and potted up once three or four leaves have formed
Begonia, cane-stemmed species	Tender perennials grown for flowers and foliage	Take softwood cuttings (see p44) in spring or summer; or use leaf slashing (see p68) or leaf squares (see p70) in spring or summer
Begonia, rhizomatous species including Rex-cultorum Group	Tender perennials grown for foliage	Divide rhizomes (see p94) in spring or summer; or use leaf slashing (see p68) or leaf squares (see p70) in spring or summer

INDOOR PLANTS FOR FLOWERS AND FOLIAGE – *BEGONIA* TO *CISSUS*

Plant name	Plant type	Propagation method
Begonia Tuberhybrida Group	Tuberous, multiflora and pendulous tender perennials grown for flowers	*Take basal cuttings (see p44) in spring or softwood cuttings (see p44) in spring or summer; or sow seed in containers (see p126) at 21°C (70°F) in early spring*
Browallia (bush violet)	Tender perennials usually grown as annuals for flowers	*Sow seed in containers (see p126) at 18°C (64°F) in early spring for summer flowering and late summer for winter to spring flowering*
Calathea	Tender perennials grown for foliage	*Divide (see p98) in late spring*
Campanula isophylla (star-of-Bethlehem)	Tender perennial grown for flowers	*Take softwood cuttings (see p44) in early spring and root with bottom heat; or divide (see p98) in spring*
Capsicum annuum (chilli pepper)	Tender annual grown for fruits	*Sow seed in containers (see p126) at 21°C (70°F) in early spring*
Catharanthus (rose periwinkle)	Tender perennials grown for flowers	*Take softwood cuttings (see p44) in late spring; or sow seed in pots (see p126) at 13–18°C (55–64°F) in early spring*
Chamaedorea	Tender palms	*Sow seed (see p126) at 25–27°C (77–81°F) in spring*
Chamaerops humilis (fan palm)	Half-hardy palm	*Separate suckers (see p75) from established plants in late spring; or sow seed (see p126) at 22–24°C (72–75°F)*
Chlorophytum comosum (spider plant)	Tender perennial grown for foliage	*Root plantlets (see p100) at any time; or divide (see p98) between spring and autumn*
Cissus (kangaroo vine)	Tender climbers grown for foliage	*Take hardwood (see p50) or greenwood (see p47) cuttings in summer*

*Flamingo flower (*Anthurium*) can be propagated by dividing plants in winter.*

Similarly, Calathea can be divided in spring to produce more plants.

Echinopsis, like most cacti, can be propagated from offsets or seed.

INDOOR PLANTS FOR FLOWERS AND FOLIAGE – *CITRUS* TO *FICUS*

Plant name	Plant type	Propagation method
Citrus (orange, lemon)	Tender trees and shrubs grown for foliage, flowers and fruit	Take semiripe cuttings (see p48) in summer
Clivia	Tender perennials grown for flowers	Divide (see p98) in late winter or early spring
Codiaeum (croton)	Tender woody-based perennials grown for foliage	Air layer (see p107) in spring; or take softwood cuttings (see p44) in summer and root with bottom heat
Cordyline, tender species	Tender tree-like perennials grown for foliage	Separate well-rooted suckers (see p75) in spring; take leaf-bud cuttings (see p58) at any time; or take root cuttings (see p74) at any time
Crassula	Succulents grown for foliage	Take leaf cuttings (see p60) or softwood cuttings (see p44) in spring or summer
Cyclamen persicum	Tender perennial grown for flowers	Soak seed for 24 hours and rinse well, then sow in containers (see p126) at 12–15°C (54–59°F) from late summer to midspring according to cultivar
Cymbidium	Tender orchids	Divide (see p98) in early or midspring when overflowing the pot
Cyperus alternifolius (umbrella plant)	Tender perennial grown for foliage	Divide (see p98) in spring; sow seed (see p126) at 18–21°C (64–70°F) in spring; or encourage plantlets by detaching a leafy flowerhead and putting it upside down in water
Dieffenbachia (dumb cane)	Tender perennials grown for foliage	Take softwood cuttings (see p44) in spring or summer; or air layer (see p107) at any time
Dracaena (dragon tree)	Tender shrubs and trees grown for foliage	Take semiripe cuttings (see p48) or leafless stem sections in summer and root with bottom heat
Echinopsis	Cacti	Separate offsets (see p100) in spring or summer; or sow seed in containers (see p126) at 21°C (70°F) in spring
Euphorbia milii (crown-of-thorns)	Tender shrub grown for flower-like bracts	Take softwood cuttings (see p44) in spring or early summer and root with bottom heat
Euphorbia pulcherrima (poinsettia)	Tender shrub grown for flower-like bracts	Take softwood cuttings (see p44) in spring or early summer and root with bottom heat
Exacum affine (Persian violet)	Tender perennial grown as an annual for fragrant flowers	Sow seed in containers (see p126) at 18°C (64°F) in early spring
Ficus (fig), tender tree species	Tender trees grown for foliage	Take softwood cuttings (see p44) or leaf-bud cuttings (see p58) and root with bottom heat in spring and summer. Air layer (see p107) F. benjamina, F. elastica and F. lyrata

INDOOR PLANTS FOR FLOWERS AND FOLIAGE – *FITTONIA* TO *PAPHIOPEDILUM*

Plant name	Plant type	Propagation method
Fittonia	Tender perennials grown for foliage	Layer (see p104) in spring or summer; or take softwood cuttings (see p44) in spring
Fuchsia	Hardy to tender perennials grown for flowers	Take softwood cuttings (see p44) in spring or semiripe cuttings (see p48) in summer
Gardenia	Tender shrubs grown for fragrant flowers	Take greenwood cuttings (see p47) in late spring or early summer or semiripe cuttings (see p48) in late summer
Guzmania	Bromeliads grown for foliage	Separate offsets (see p100) in midspring
Hippeastrum (amaryllis)	Tender perennials grown for flowers	Separate bulblets (see p86) in autumn
Howea fosteriana (kentia palm)	Tender palm	Sow seed in containers (see p126) at 26°C (79°F) as soon as ripe
Hoya (wax plant)	Tender climbers grown for fragrant flowers	Layer (see p104) in spring or summer; or take semiripe cuttings (see p48) in late summer and root with bottom heat
Hypoestes (polka-dot plant)	Tender shrubby perennials grown for foliage	Sow seed in pots (see p126) at 15–18°C (59–64°F) in spring; take softwood cuttings (see p44) in late spring or semiripe cuttings (see p48) in summer
Impatiens (busy Lizzie)	Tender perennials grown for flowers and, in some cultivars, foliage	Take softwood cuttings (see p44) in spring or early summer
Kalanchoe	Succulents, some grown for flowers, others for foliage	Take softwood cuttings (see p44) in spring or summer; remove offsets (see p100) or plantlets at any time; or sow seed in pots (see p126) at 21°C (70°F) in early spring
Maranta leuconeura (prayer plant)	Tender perennial grown for foliage	Divide (see p98) in spring; or take basal softwood cuttings (see p44) in spring and root with bottom heat
Monstera deliciosa (Swiss-cheese plant)	Tender climber grown for foliage	Layer (see p98) in autumn; or root softwood (see p44) or leaf-bud (see p58) cuttings with bottom heat in summer
Nephrolepis (sword fern)	Tender evergreen ferns	Separate rooted runners (see p100) in late summer or early spring
Nertera granadensis (bead plant)	Tender perennial grown for berries	Divide (see p98) in spring; or sow seed in containers (see p126) at 13–16°C (55–61°F) in spring
Odontoglossum	Tender orchids	Divide (see p98) when overflowing the pot in early spring or late summer
Paphiopedilum (slipper orchid)	Tender orchids	Separate offsets (see p100), or take softwood cuttings (see p44) in spring

INDOOR PLANTS FOR FLOWERS AND FOLIAGE – *PELARGONIUM* TO *ROSA*

Plant name	Plant type	Propagation method
Pelargonium (geranium)	Tender perennials grown for flowers and foliage	Take softwood cuttings (see p44) in spring, late summer or early autumn
Peperomia	Tender perennials grown for foliage	Remove offsets (see p100); or take leaf-stalk (see p62), leaf-bud (see p58) or softwood (see p44) cuttings from midspring to early autumn
Pericallis × hybrida (cineraria)	Tender perennial grown as annual for flowers	Sow seed in pots (see p126) at 13–18°C (55–64°F) from spring to midsummer; or take semiripe cuttings (see p48) in summer
Phalaenopsis (moth orchid)	Tender orchids	Separate offsets (see p100) in spring; or take softwood cuttings (see p44) in spring
Philodendron	Tender climbers grown for foliage	Layer (see p104) or air layer (see p107) in spring; or take softwood (see p44) or leaf-bud (see p58) cuttings in summer
Pilea	Tender perennials grown for foliage	Divide (see p98) in spring; or take softwood cuttings (see p44) in spring
Plectranthus	Tender perennials grown for foliage	Take softwood cuttings (see p44) at any time; or divide (see p98) from mid- to late spring
Radermachera	Tender trees grown for foliage	Take softwood cuttings (see p44) in summer
Rhododendron simsii (Indian azalea)	Tender shrub grown for flowers	Take hardwood cuttings (see p50) from midsummer to early autumn and root with bottom heat
Rosa, miniature varieties	Hardy shrub grown for flowers	Take hardwood cuttings (see p50) in late summer or early autumn

Propagate wax plant (Hoya) by layering or take semiripe cuttings.

Bead plant (Nertera granadensis) can be propagated by division or by sowing seed.

African violet (Saintpaulia) is easily propagated from leaf-stalk cuttings.

INDOOR PLANTS FOR FLOWERS AND FOLIAGE – *SAINTPAULIA* TO *YUCCA*

Plant name	Plant type	Propagation method
Saintpaulia (African violet)	Tender perennials grown for flowers	Take leaf-stalk cuttings (see p62) in summer
Sansevieria (mother-in-law's tongue)	Tender perennials grown for foliage	Separate offsets (see p100) in spring; divide (see p98) in spring; or take monocot leaf cuttings (see p72) from spring to autumn and root with bottom heat
Schefflera	Tender trees and shrubs grown for foliage	Take semiripe cuttings (see p48) in summer and root with bottom heat; or air layer (see p107) in spring
Schlumbergera (Christmas cactus)	Tender forest cactus grown for flowers	Take greenwood cuttings (see p47) in spring or early summer
Sinningia speciosa (gloxinia)	Tender perennial grown for flowers	Take softwood cuttings (see p44) in late spring or early summer; divide tubers (see p92) in spring; or take leaf-stalk cuttings (see p62) or midrib cuttings (see p64) in spring or summer
Solanum pseudocapsicum (winter cherry)	Tender shrub grown as an annual for fruit	Sow seed in containers (see p126) at 18–20°C (64–68°F) in spring
Solenostemon scutellarioides (coleus)	Tender perennial grown for foliage	Take softwood cuttings (see p44) in spring or summer; or surface sow seed in containers (see p126) at 22–24°C (72–75°F) in early spring
Spathiphyllum (peace lily)	Perennials grown for flowers and foliage	Divide (see p98) in winter, or immediately after flowering
Stephanotis floribunda (bridal wreath)	Tender climber grown for fragrant flowers	Root semiripe cuttings (see p48) with bottom heat in summer
Streptocarpus (Cape primrose)	Tender perennials grown for flowers	Divide (see p98); or take midrib (see p64) or lateral vein (see p66) cuttings in spring or early summer. Take softwood cuttings (see p44) of bushy or trailing species in spring and root with bottom heat
Syngonium (goosefoot plant)	Tender climbers grown for foliage	Take softwood (see p44) or leaf-bud (see p58) cuttings in summer and root with bottom heat
Tillandsia (air plant)	Bromeliads grown for foliage and flowers	Separate offsets (see p100) in spring
Tradescantia, tender species	Tender perennials, many trailing, grown for foliage	Take softwood cuttings (see p44) of trailers at any time. Divide clump-forming species (see p98) in spring or autumn
Yucca eliphantipes (spineless yucca)	Tender perennial grown for palm-like habit	Separate suckers (see p75); or take hardwood cuttings (see p50) in spring

• GLOSSARY •

Adventitious Formed accidentally, casually or in an unusual anatomical position. A shoot or root, for example, that develops from tissue that would not normally produce that organ.

Air layering see p107.

Annual A plant that completes its life cycle in one season.

Apical bud see p39.

Axillary bud see p39.

Biennial A plant that lives for two years, growing vegetatively in the first year and flowering and setting seed in the second.

Bottom heat Warmth provided below the potting compost or the container, usually by an electric propagating unit.

Budding The process of grafting a bud onto a rootstock.

Bulbil A small, bulb-like structure, especially in the axil of a leaf or at the base of a stem, that may form a new plant.

Bulblet A small or offset bulb.

Callus The protective, often corky, wound tissue that develops on a damaged or exposed surface.

Cambium A simple plant cell or tissue from which conductive tissues or cork develop, resulting in secondary thickening in woody plants.

Cormel or cormlet A small corm.

Crown The part of a plant at or around ground level that produces stems.

Cutting A separated piece of stem, root or leaf that has been prepared for propagation of a new plant.

Deciduous Used to describe a plant that drops its leaves in autumn.

Dibber A tool used to make a small hole in which to plant a cutting, seedling or small plant.

Drill A shallow groove or furrow made in the soil or compost in which to sow seed.

Dropping see p112.

Ericaceous Used to describe plants that are lime hating and need an acidic (lime-free) soil or compost.

Etiolated Used to describe a pale and thin plant that is drawn up as a result of insufficient amounts of light.

Eye A growth bud.

F1 variety/hybrid A plant or seeds that have been bred under strict conditions to create a variety that is uniform, vigorous and with highly desirable attributes.

Graft union The point at which a rootstock and scion have been joined by grafting.

Greenwood cutting see p47.

Hardening off The process of acclimatising tender or half-hardy plants raised with warmth indoors to outdoor conditions by gradual exposure to lower temperatures.

Heel cutting see p56.

Internode A length of stem between two nodes.

Lateral shoot A sideshoot.

Leaf axil see p39.

Mallet cutting see p57.

Node A place where a leaf joins the stem.

Parent plant see p6.

Perennial A plant living for more than two years.

Petiole A leaf stalk.

Photosynthesis The process by which green plants use sunlight to produce foods from carbon dioxide and water.

Prick out To move a seedling or young plant from one place to another: for example, a seedling from a small to a larger pot or out into the garden.

Propagator A heated or unheated, glass- or plastic-covered, box-like structure for raising seedlings or cuttings.

Rootstock A plant onto which another variety/cultivar is grafted.

Scion The part of the plant (single bud or stem) that is grafted onto a rootstock and produces the topgrowth of the new plant.

Scooping and scoring see p89.

Station sowing To sow seed individually at a predetermined spacing.

Stratification The process of breaking seed dormancy by mimicking the conditions naturally occurring in the environment.

Transplant To move a seedling or young plant from one place to another: for example, a seedling from a small to a larger pot or out into the garden. Also [noun] a plant being so moved.

Vegetative Used to describe reproduction/propagation by asexual means, as opposed to sexual propagation (seed production). Also used to describe plant growth not associated with flowering.

· INDEX ·

A
air layering 107–9
alpines 129
annuals 8, 116, 118, 48–52
aphids 34
apical-wedge grafting 140–1

B
basal cuttings 44, 49
basal rot 35
bedding 38, 44, 116, 148–52
biennials 148–52
biological controls 34
blocks 23
broadcasting 119
bulbs 84–5
 bulb scales 87, 88–9
 bulbils 86
 bulblets 86–8
 scooping 88, 89
 scoring 89
burying 113

C
canes 18, 20
Cape primrose 61, 64, 66, 67
celltrays 23–4
chip-budding 144–5
cleft grafting 140
clematis 58, 108
climbers 38, 59
 serpentine layering 108
cold frames 15–16
compost tampers 18, 19
composts 26–7
conifers 38
containers 22–5
 hardwood cuttings 50
 reusing 25
 sowing seed 126–9
 tidemarks 33
cormels (cormlets) 91
corms 84, 85, 90–1
cultivars 12
cuttings 6, 7, 9
 compost 27
 cuttings boards 18, 21

D
diseases 12, 32–3, 35
division 98–9
 corms 90
 fibrous crowns 99
 fleshy crowns 98
 tuberous root cuttings 83
 tubers 93
dogwood (Cornus) 52, 53, 108
dormancy 122, 129
 cutting or chipping 122
 hot water 122, 125
 scarification 124
 scraping 122
 stratification 125
drilling 119
dropping 112–13

E
equipment 18–21
evergreen cuttings 54–5
 winter 54

F
F1 hybrids 12, 119, 120
fleece 18, 21
French layering 108
fruit 38, 50, 75, 120
 grafting 136, 138–9
 tip layering 110–11

G
genetics 119
germination 12, 126, 127, 129
 problems 132
glasshouses 14–15, 29
 shading 14
grafting 7, 136–7
 apical-wedge grafting 140–1
 chip-budding 144–5
 side-wedge grafting 142–3
 whip & tongue 138–9
greenwood cuttings 42, 47

H
hairy leaves 71
hand sprayers/misters 18, 20
hardening off 16
hardwood cuttings 9, 42, 50–3
 containers 50
 hollow-stemmed cuttings 53
 single stems 53
 slow-rooting plants 50
 soft-pith cuttings 53
hedges 7
heel cuttings 56–7
herbs 38
hollow-stemmed cuttings 53
home-made composts 27
hooping leaf cuttings 73
hormone rooting
 products 18, 20, 29
hygiene 32–3

I
indoor plants 38, 44, 182–7
 air layering 107–9
 lateral vein cuttings 66–7
 leaf cuttings 60, 61
 leaf slashing 68–9
 leaf squares 70–1
 leaf-stalk cuttings 62–3
 midrib cuttings 64–5
 monocot leaf cuttings 72–3
internodal cuttings 43
irises 94, 95

K
knives 18–19, 20
 cutting hardwood 20
 sharpening 19, 20

L
labelling 18, 19
lateral vein cuttings 66–7
layering 104–5
 air layering 107–9
 French layering 108
 serpentine layering 108
 simple layering 104–6
 stem girdling 109
 tip layering 110–11
leaf cuttings 6, 60
 making cuttings 60
 suitable plants 61

whole leaves 60
leaf slashing 68–9
leaf squares 70–1
leaf-bud cuttings 58–9
   double leaf-bud cuttings 58
   vine-eye cuttings 59
leaf-stalk cuttings 62–3
loam-free compost 27

M
mallet cuttings 57
micropropagation 9
midrib cuttings 64–5
mist propagation 17
monocot leaf cuttings 72–3
mounding up 113

N
nodal cuttings 43

O
offsets 100–1

P
parent plants 6
   preparation 76–7
pellets 23
perennials 38, 74, 116, 153–66
   basal cuttings 44
   division 98–9
pests 34–5
Plant Breeders' Rights (PBR) 8
plant material 12–13
plastic covers 17, 18, 21
polythene bags 17, 19–20
potatoes 92
pots 22
   disposable pots 23
   paper pots 24
potting compost 27
propagation 6–9
   environmental control 14–17
   success rate 7
propagators 15, 16–17

R
recycling 25
rhizomes 84, 85, 94–5
   crown rhizomes 95

dividing 94
root cuttings 6, 9, 74
   obtaining material 78
   optimum size 79
   preparation 76–7, 79
   starting 80–1
   tuberous cuttings 82–3
   woody shrubs 75
root disturbance 81
roses 38, 50, 75
runners 101

S
scarification 124
sciarid flies 34, 35
secateurs 18, 19
   secateur types 21
seedlings 130–1
   problems 132–3
seeds 6, 7, 8
   breaking dormancy 122–5
   collecting 120–1
   compost 27
   germinating seeds 131
   natural food reserves 31
   seed morphology 116
   self-seeding 8–9
   storing 121
   viability test 1212
seedtrays 23–4
semiripe cuttings 9, 42, 48
   basal cuttings 49
serpentine layering 108
shrubs 38, 44, 50, 74, 116
   basal cuttings 49
   division 98
   grafting 136, 138–9, 140–1, 142–3
   greenwood cuttings 47
   root cuttings 75
side-wedge grafting 142–3
simple layering 104–6
single-stemmed plants 53
slugs 35
soft-pith cuttings 53
softwood cuttings 9, 42, 44–7
   basal cuttings 44

soil blockers 24
sowing 6, 7, 8
   containers 126–9
   outdoor sowing 116–19
   sowing techniques 129
   troubleshooter 132–3
spider mites 34, 35
stem cuttings 6, 13, 40
   improving success 38
   old stems 40
   parts of a stem 39, 85
   root development 29
   types 42
   when to take 43
   where to cut 43
   winter timing 43
stem girdling 109
stratification 125
suckers 75, 98

T
thin roots 81
tissue culture 9
tools 18–21
   basic toolkit 18
   sterilising 19
trees 38, 44, 50, 75, 116
   grafting 136–7, 138–9, 140–1, 142–3
tuberous root cuttings 82–3
tubers 84, 85, 92–3
   tubercles 93
   types 82

V
vine-eye cuttings 59

W
watering 30–1
watering cans 18, 20
whip & tongue grafting 138–9
whitefly 34, 35
windowsill propagators 16
woody plants 75, 167–81
   dropping 112–13
   layering 104–11
wrinkled leaves 71

• PICTURE CREDITS •

Note The acknowledgements below appear in source order.

Alamy John Glover 134–135; katewarn images_floral 114–115; Tim Gainey 117

Fotolia Alison Bowden 178 left; apple1 151 centre; Exsodus 186 left; Gabriella88 159 left; glasscutter 162 centre; Kerioak 74; laurent007 175 centre; Mario 178 centre; Nathalie Moscowskaya 186 centre; Ringelblume 151 left; Tetiana Zbrodko 137; Thorsten Schier 167 centre

GAP Photos BBC Magazines Ltd 41; Clive Nichols/Cambridge Botanic Garden 109; Friedrich Strauss 10–11, 45; Jonathan Buckley/demonstrated by Carol Klein 118; Keith Burdett 31; Mark Winwood 102–103; Rachel Warne 146–147; Richard Bloom 2, 13

Garden World Images Andrea Jones 9; Glenn Harper 15; Jacqui Dracup 63, 64; Jenny Lilly 7; John Swithinbank 25, 33; Liz Cole 32; Liz Every 183 centre; MAP/Nicole et Patrick Mioulane 34; Mein Schöner Garten 96–97; N+R Colborn 36–37; Nicole et Patrick Mioulane 175 right; Rita Coates 67; Steffen Hauser 140; Trevor Sims 93

Octopus Publishing Group 151 right, 154 all, 159 centre and right, 162 left and right, 167 left and right, 170 all, 175 left, 178 right, 183 left and right, 186 right

Photolibrary Garden Photo World/David C Phillips 143; Garden Picture Library/Friedrich Strauss 106, /Mark Bolton 52, /Mark Winwood 94, 111; Moodboard RF 28